AUTOMATIC LEARNING TECHNIQUES IN POWER SYSTEMS

THE KLUWER INTERNATIONAL SERIES IN ENGINEERING AND COMPUTER SCIENCE

Power Electronics and Power Systems

Consulting Editors

Thomas A. Lipo and M. A. Pai

Other books in the series:

ENERGY FUNCTION ANALYSIS FOR POWER SYSTEM STABILITY, M. A. Pai, ISBN: 0-7923-9035-0
ELECTROMAGNETIC MODELLING OF POWER ELECTRONIC CONVERTERS, J. A. Ferreira, ISBN: 0-7923-9034-2
MODERN POWER SYSTEMS CONTROL AND OPERATION, A. S. Debs, ISBN: 0-89838-265-3
RELIABILITY ASSESSMENT OF LARGE ELECTRIC POWER SYSTEMS, R. Billington, R. N. Allan, ISBN: 0-89838-266-1
SPOT PRICING OF ELECTRICITY, F. C. Schweppe, M. C. Caramanis, R. D. Tabors, R. E. Bohn, ISBN: 0-89838-260-2
INDUSTRIAL ENERGY MANAGEMENT: *Principles and Applications*, Giovanni Petrecca, ISBN: 0-7923-9305-8
THE FIELD ORIENTATION PRINCIPLE IN CONTROL OF INDUCTION MOTORS, Andrzej M. Trzynadlowski, ISBN: 0-7923-9420-8
FINITE ELEMENT ANALYSIS OF ELECTRICAL MACHINES, S. J. Salon, ISBN: 0-7923-9594-8

AUTOMATIC LEARNING TECHNIQUES IN POWER SYSTEMS

Louis A. Wehenkel
National Fund for Scientific Research
University of Liège - Institut Montefiore
Sart-Tilman B28, B-4000 Liège, Belgium

Kluwer Academic Publishers
Boston/London/Dordrecht

Distributors for North America:
Kluwer Academic Publishers
101 Philip Drive
Assinippi Park
Norwell, Massachusetts 02061 USA

Distributors for all other countries:
Kluwer Academic Publishers Group
Distribution Centre
Post Office Box 322
3300 AH Dordrecht, THE NETHERLANDS

Library of Congress Cataloging-in-Publication Data

A C.I.P. Catalogue record for this book is available
from the Library of Congress.

Printed on acid-free paper.

Printed in the United States of America

Most of the illustrations of automatic learning methods given in this book have been realized with a data mining software called ***GTDIDT***.
GTDIDT is written in Common Lisp and was developed at the University of Liège for research. It was used in all the large scale power system applications described in this book.
In order to enable the reader to try out the methods presented in this book, a free of charge demo version of ***GTDIDT***, based on ***GNU Common Lisp*** from the ***Free Software Foundation***, is available from the WWW home page of the author at the URL

`"http://www.montefiore.ulg.ac.be/~lwh/"`

It contains a small transient stability data base (one-machine infinite bus system) as well as many of the data mining algorithms described in the book.

Contents

List of Figures

List of Tables

Preface

A book should have either intelligibility or correctness; to combine the two is impossible.
—Bertrand Russel (1872-1970)

Automatic learning is a complex, essentially multidisciplinary field of research and development, involving theoretical and applied methods from statistics, computer science, artificial intelligence, biology and psychology. Its applications to engineering problems, such as those encountered in electric power systems, are therefore challenging, while extremely promising. More and more data become available, collected from the field by systematic archiving, or generated through computer based simulation. To face this explosion of data, automatic learning can provide systematic approaches, without which only poor use could be made of growing amounts of data and computing power.

One of the objectives of this book is to present in a consistent way a representative subset of automatic learning methods - basic and more sophisticated ones - available from statistics (both classical and modern), and from artificial intelligence (both hard and soft computing). Another one is to discuss appropriate methodologies for combining these methods to make the best use of available data in the context of real-life problems.

The book thus considers automatic learning from the application perspective, and essentially aims at encouraging research and especially applications. Indeed, during the last 15 years, pattern recognition, expert systems, artificial neural nets, fuzzy systems, evolutionary programming, and other artificial intelligence paradigms, have yielded an impressive number of publications in the power system community, but only very few real applications. Too many of these publications were seeking for novelty, sometimes overlooking the fact that existing classical techniques could do the job as well. In this book I have therefore selected a subset of automatic learning methods mainly on the ground of their functionalities and potentials in practice, rather than based on novelty, or intellectual appeal.

To avoid duplication with a growing literature on automatic learning per se, I have opted to describe a tool-box of methods which I found to be practically useful in the context of my research, and complementary to each other. While, without doubt, new, more powerful methods will appear in the future, as research goes on, I am convinced that the tool-box approach will remain. Indeed, no automatic learning method can claim to be universal. Besides, I also believe that many of the methods, sometimes very simple ones, described in this book, will remain competitive in the future.

Due to space limitations, many interesting topics in automatic learning could not be included in the book. In particular, learning theories and interesting concepts from multi-strategy learning systems are only mentioned. Furthermore, the book only considers data intensive methods of supervised and unsupervised learning, skipping subjects like knowledge intensive learning, machine discovery, or learning by analogy.

On the other hand, to convince engineers of the usefulness of automatic learning, and also to make the presentation more lively, most of the methods are illustrated on real problems. Some of the questions raised by these practical applications (e.g. how to exploit temporal data in automatic learning) are not yet solved in a satisfactory way. Hence, I hope that the book will contribute to foster research and new ideas to solve these problems.

An important part of the book is dedicated to the practical application of automatic learning to power systems. Since, in many of these applications, data bases are obtained from simulations, I discuss in detail data base generation and validation methodologies. Furthermore, many different ways of applying automatic learning are illustrated and the complementary features of different methods are stressed. Power system problems to which automatic learning can be applied are screened and the complementarity of automatic learning in general with respect to analytical methods and numerical simulation are investigated.

About the title

The book is entitled "Automatic learning techniques in power systems" suggesting a very broad application scope.

However, the reader will notice that the core of the applications discussed concerns security assessment related subjects. This is so for several good and less good reasons. First of all, security assessment is probably the most versatile field of application. It has led to a quite rich literature and an advanced application stage. Moreover, with the present day evolution of power systems, security assessment will become more and more challenging, and there is a true risk that power system reliability decreases if no proper methodologies are put in place soon enough. I believe that automatic learning provides real possibilities in this context. The second, maybe less good reason for concentrating on this topic is that it has been my essential field of research during 12 years. This allowed me to draw many examples from real application studies, and hopefully to make my message become more credible.

Nevertheless, I am convinced that many other interesting opportunities for applying automatic learning to power systems exist. I would like to encourage power system engineers to look at automatic learning in a sufficiently broad perspective, including methods from classical statistics, symbolic artificial intelligence and neural networks. The title suggests that many of the techniques and methodologies discussed in this book can be carried over to these other applications.

Outline of the book

The book is organized into an introduction and three parts.

The Introduction first considers automatic learning and its application to power systems from a historical point of view, then provides an intuitive introduction to supervised and unsupervised learning. A simple transient stability assessment problem is used to describe decision tree induction, multilayer perceptrons, nearest neighbor and clustering methods, in order to highlight key concepts in automatic learning and functionalities of various classes of methods.

Part I focuses on theoretical aspects of AL, and describes the main classes of such methods. Chapter 2 defines supervised and unsupervised learning as search problems, and introduces notation and main unifying concepts used throughout the rest of Part I. The following 4 chapters screen various types of automatic learning methods. We start, in Chapter 3, by presenting statistical methods, because they essentially provide the foundations of automatic learning, and because most readers are already familiar with some of them. In Chapter 4, artificial neural network based methods are described, insisting on their similarities to modern statistical ones. Chapter 5 introduces symbolic methods from machine learning, mainly providing a detailed description of decision tree induction. Finally, Chapter 6 describes data preprocessing techniques of general use, discusses complementary roles of various AL algorithms, and suggests interesting ways to combine them into more powerful hybrid methods.

Part I assumes that the reader has some basic background in probability theory, linear algebra, calculus and function optimization. However, detailed mathematical derivations have been reduced as much as possible, and many real-life examples are used to illustrate ideas. As a guide for further reading, a short list of selected references is provided at the end of Chapters 2 to 5.

Part II concentrates on security assessment applications. Chapter 7 introduces a general framework for the application of AL to security assessment. Chapter 8 screens security assessment problems, discussing AL potentials in various practical contexts. Chapter 9 provides a detailed discussion about data base generation techniques, which are paramount for the successful application of AL to security assessment. These are illustrated in Chapter 10 on a sample of detailed real-life case studies. Finally, Chapters 11 and 12 discuss respectively the added value of AL with respect to existing analytical approaches and the future orientations, where the field is likely to go. Only elementary background in power system security and stability is assumed here.

Part III contains a single chapter which provides a brief survey of the main other application fields of AL in power systems. Without aiming at exhaustiveness, my purpose here is to suggest that many problems could benefit from automatic learning.

The bibliography at the end of the book mainly collects reference texts in automatic learning and publications of its applications in power systems.

The most frequently used mathematical notations are collated at the beginning of the book, while, at the end, an index is provided together with a glossary of acronyms used in the text. The (foot)notes of each chapter are collected at the end of the chapter.

The material in Part I gives a broad introduction to the main automatic learning methods; it is used as the core text of a course on "Applied automatic learning" given to undergraduate and graduate students in Engineering and Mathematics at the University of Liège.

The reader who is more interested in the applications to power systems should read the Introduction before jumping to Part II.

Chapter 10, which gives many details of some large scale studies can be skipped in a first pass.

LOUIS WEHENKEL

Liège, Belgium

This book is dedicated to my
colleagues and friends who
contributed to my research
during the last 12 years and to
my family who has accepted
to live with a researcher.

Acknowledgments

Without the collaboration of Professor Mania Pavella (University of Liège) the monograph would not exist. Since the beginning of my research, she always provided the right balance of criticism and enthusiastic feedback and, thanks to her, some of my colleagues worked together with me on this subject. She also helped a lot to improve the manuscript.

Much of the material presented in this book was obtained through teamwork with colleagues at the University of Liège and research collaborations with engineers from Electricité de France and Hydro-Québec. While they are too numerous to mention individually, their contributions have been paramount to the success of this book. Special thanks go to Yannick Jacquemart and Cyril Lebrevelec (Electricité de France). Both of them collaborated with me during 16 months at the University of Liège. Yannick Jacquemart provided part of the material presented in Chapter 10.

MLP simulations were carried out with SIRENE, developed at the University of Liège by Michel Fombellida. Kohonen feature maps were based on SOM_PAK, developed at the University of Helsinki by the team of Teuvo Kohonen. I thank both of them for making their software available.

Part of Chapter 1 and Part II are adapted from a Tutorial course I gave with Yannick Jacquemart at the IEEE Power Industry Computer Application Conference, in May 1997. The sponsoring of the PICA Tutorial Committee is greatly appreciated.

I would like to thank Professor Pai, from the University of Illinois, who proposed me to write the monograph and provided enthusiastic and insightful advises. My thanks go also to Mr Alex Greene from Kluwer Academic Publishers, who believed in the project from the beginning and provided the appropriate pressure to make it become reality.

Last, but not least, I am pleased to acknowledge the help of Ms Cristina Olaru (PhD student at the University of Liège) who contributed a lot to improve clarity and consistency throughout the text.

Notation

Below we provide a short summary only of the general most frequently used notation. A glossary defining acronyms and other frequently used symbols is provided at the end of the book.

$\boldsymbol{x}, \boldsymbol{y}, \ldots$	lowercase boldface letters denote (column) vectors
$A, B \ldots$	uppercase letters are used to denote matrices
$\boldsymbol{x}^T, A^T$	transpose of $\boldsymbol{x}$, A
$\boldsymbol{U}$	universe of possible objects o
$\boldsymbol{X}, \boldsymbol{Y}, \ldots$	uppercase boldface letters denote subsets of $\boldsymbol{U}$
$\boldsymbol{X}^{\boldsymbol{X}}$	the set of all subsets of $\boldsymbol{X}$ (power set)
$\|\boldsymbol{X}\|$	the size (cardinality) of a set $\boldsymbol{X}$
$\neg\boldsymbol{X}$	the subset of objects not belonging to $\boldsymbol{X}$
$\boldsymbol{X} \cap \boldsymbol{Y}$	intersection of sets
$\boldsymbol{X} \cup \boldsymbol{Y}$	union of sets
$P = \{\boldsymbol{P}_1, \ldots, \boldsymbol{P}_p\}$	partition of $\boldsymbol{U}$
$f(\cdot), g(\cdot), \ldots$	functions defined on $\boldsymbol{U}$
$f(o), g(o), \ldots$	function values of object o
$a(\cdot)$	an attribute (a function of objects)
n	the number of different attributes
$\boldsymbol{a}(o)$	n-dimensional vector of attribute values of o
$c(\cdot)$	a classification function
$d(\cdot)$	a decision rule (a guess of a learning algorithm for $c(\cdot)$)
m	the number of classes, i.e. of different values of $c(\cdot)$
$\{c_1, \ldots, c_m\}$	the set of possible values of $c(\cdot)$
$\boldsymbol{C}_i$	the set of all objects such that $c(o) = c_i$
$C = \{\boldsymbol{C}_1, \ldots, \boldsymbol{C}_m\}$	the goal partition of $\boldsymbol{U}$

$y(\cdot)$ (resp. $\boldsymbol{y}(\cdot)$)	a real valued scalar (resp. vector) regression function
$r(\cdot)$ (resp. $\boldsymbol{r}(\cdot)$)	a regression model (a guess of a learning algorithm for $y(\cdot)$)
$\mathcal{F}, \mathcal{G}, \mathcal{D}, \mathcal{R} \ldots$	calligraphic letters denote sets of functions
$P(\boldsymbol{X})$	the prior probability of $\boldsymbol{X}$
$P(\boldsymbol{X}\|\boldsymbol{Y})$	the posterior probability of $\boldsymbol{X}$ with respect to $\boldsymbol{Y}$
δ_{ij}	the Kronecker symbol ($\delta_{ij} = 1$ if $i = j$, 0 otherwise)
$f'(\cdot)$	the derivative of a function $f(\cdot)$ defined on R
inf	lower bound of a set of numbers
R^n	Euclidean space of n-dimensional vectors of real numbers.

AUTOMATIC LEARNING TECHNIQUES IN POWER SYSTEMS

1 INTRODUCTION

1.1 HISTORICAL PERSPECTIVE ON AUTOMATIC LEARNING

The term Automatic Learning (AL) is nowadays used to denote a highly multidisciplinary research field and set of methods to extract high level synthetic information (knowledge) from data bases containing large amounts of low level data. The researchers in the field are statisticians and computer scientists, but also psychologists and neuro-physiologists. The latter's aim is mainly to understand human learning abilities and they use automatic learning algorithms as models of natural learning. The former concentrate more on the computational properties of such algorithms, mainly to enable them to solve engineering problems. AL encompasses statistical data analysis and modeling, artificial neural networks, and symbolic machine learning in artificial intelligence. Related work in statistics dates back to Laplace [Lap10] and Gauss [Gau26]. In the field of artificial neural networks, the early attempts were in the 1940's [MP43], while work in symbolic machine learning emerged only in the mid sixties [HMS66].

In the last two decades, automatic learning has progressed along many lines, in terms of theoretical understanding (see e.g. [Wol94]) and actual applications in diverse areas. Probably the main reason for the important breakthrough was the tremendous increase in computing powers. This makes possible the application of the often very compute intensive automatic learning algorithms to practical large scale problems. Conversely, automatic learning algorithms allow one to make better use of existing computing power by exploiting more systematically the information contained in data bases. The availability of very large scale data bases waiting for being properly exploited is thus a very strong, though recent motivation fostering research in automatic learning.

Nowadays, automatic learning methods thus receive routine applications, for example in medical diagnosis, in character recognition and image processing, as well as in financial and marketing problems.

Recently, the term Knowledge Discovery from Data bases (KDD) was coined to denote the emerging research and development field aiming at developing methodologies and software environments for large scale applications of automatic learning [FPSSU96]. Hence, this book is mainly concerned with the application of KDD to power system problems.

Application to power systems security assessment. In the context of power systems security assessment (SA), research on automatic learning started with Pattern Recognition (PR) in the late sixties and seventies [Dy 68, PPEAK74, GEA77].

Initially the goal of this work was to improve security assessment by combining analytical system theory tools (mainly numerical simulation) with statistical information processing techniques from automatic learning. In particular, analytical tools are exploited in this scheme to screen ranges of security scenarios and to build data bases containing detailed security analysis results. Automatic learning methods are then applied to extract relevant synthetic information from these data bases in various forms, in order to figure out under which conditions a particular power system is secure, in a particular context. After a careful validation, the extracted knowledge eventually is translated into planning and operating decisions. We will see in the following chapters that this is a very flexible framework, which may be applied to a large diversity of SA problems.

With a proper methodology, the information extracted by automatic learning is indeed complementary to classical system theory methods along three dimensions : computational efficiency (in terms of response time and data requirements); interpretability (in terms of physical understanding); management of uncertainties (in terms of modeling and measurement errors).

In the early attempts, the methodology was essentially limited, on the one hand, by the small size of the security information data bases which could be managed, on the other hand, by the parametric nature of the existing pattern recognition methods, which were unable to handle properly the large scale and the nonlinear character of power system security problems. However, since the mid eighties research accelerated due to several factors.

First of all, the computing environments became powerful enough to enable the generation of rich enough security information data bases, with acceptable response times.

Second, research in automatic learning had produced new methods able to handle the complexity and non-linearity of power system security problems. In particular, artificial neural networks and machine learning methods have shown their complementary potentials, as reflected by the growing number of publications on their applications to various power system problems (e.g. [PDB85, FKCR89, SP89, ESMA+89, WVRP89a, MT91, OH91, Dil91, NG91, HCS94, RKTB94, MK95], to quote only a few ones).

The last - but not least - factor comes from the real interest shown by electric Utilities. A few years ago, some of them started ambitious research projects to apply the approach to their power system and find out whether it could comply with their practical needs. This contributed significantly to formalize the application methodology and to develop software tools, able to handle real, large scale problems. It opened also new research directions and suggested new types of applications.

Today, only a few large Utilities in North-America and Europe have actually assessed automatic learning based security assessment in the context of their specific power systems. At least one of them - Electricité de France - is presently using the methodology in real field studies. Certainly, some others will start using it in the near future.

Indeed, the fast changes which take place in the organization of power systems imply the use of more systematic approaches to dynamic security assessment in order to maintain reliability at an acceptable level. In the future, power systems will behave more erratically, and at the same time they will operate closer to their limits. Recent blackouts in the Western USA and other places around the world, have already demonstrated that present day methodologies reach their limits under such circumstances.

Other applications to power systems. Design is an area where we believe that automatic learning has great potential, especially design of robust and/or adaptive systems. The methodologies developed in security assessment can indeed be carried over to other design problems, for instance of control, measurement and protection systems.

Load forecasting emerged in the early seventies as a very popular application of automatic learning [De96, ESN96]. In this context, mainly artificial neural networks have been exploited as flexible nonlinear modeling tools. They already yielded various real-life applications.

Another broad area concerns process and equipment monitoring, leading to such applications as transformer monitoring and power plant monitoring.

Finally, load modeling, state estimation and optimization are fields where automatic learning has been proposed in the literature.

More generally speaking, due to the low cost of mass storage devices and significant improvements in measurement systems, Utilities are able to collect huge amounts of data from the field and to archive them in data warehouses. Thus, in order to exploit these data we may foresee that many new KDD applications will emerge in the future.

We will provide a brief overview of some of these applications in the last part of the monograph.

1.2 AN AUTOMATIC LEARNING TOOL-BOX

The aim of this section is to give a flavor of what automatic learning is all about before leading the reader into more involved discussions in the subsequent chapters. In particular, we will highlight the complementary nature of different methods so as to motivate a tool-box approach to automatic learning.

We first introduce the *supervised* learning problem and illustrate it by a simple power system transient stability example. Then, we describe three popular automatic learning methods, representative of three complementary classes of methods. We conclude the chapter by briefly discussing unsupervised learning. Although we aim to stay at an intuitive level, on the road we introduce key concepts and main technical aspects in automatic learning.

1.2.1 Supervised learning

At an abstract level, the generic problem of supervised learning from examples can be formulated as follows :

> *Given a set of examples (the learning set (LS)) of associated input/output pairs, derive a general rule representing the underlying input/output relationship, which may be used to explain the observed pairs and/or predict output values for any new unseen input.*

In automatic learning we use the term *attribute* to denote the parameters (or variables) used to describe the input information. The output information can be either symbolic (i.e. a *classification*) or numerical.

In the context of security assessment, an *example* would thus correspond to an operating state of a power system, or more generally to a simulated security scenario. The input attributes would be relevant parameters describing its electrical state and topology and the output could be information concerning its security, in the form of either a discrete classification (e.g. secure / insecure) or a numerical security margin. This is illustrated by the example below.

1.2.2 Illustrative transient stability problem

Figure 1.1 depicts a simple One-Machine-Infinite-Bus (OMIB) system, composed of a generator, a transformer, a load connected at the EHV (400kV) side of this transformer, and a reactance (Xinf) representing the equivalent impedance of the EHV network.

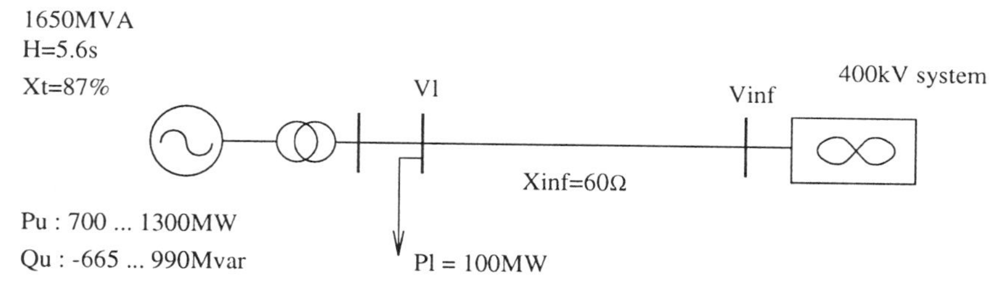

Figure 1.1. One-machine infinite bus system

The generator inertia constant H and the reactance Xt (modeling the transient direct axis reactance, the transformer reactance, and a short line connecting the generator to the EHV substation) are given in p.u. of its nominal MVA rating (1650MVA).

Let us consider that a three-phase short-circuit occurs in the EHV substation close to the generator, normally cleared (without line-tripping) after 155ms, and let us declare the system as insecure if the generator loses synchronism after this disturbance. To determine the system security degree, we will evaluate the *critical clearing time* (CCT) of this disturbance, which is the maximum time for the short-circuit to be cleared without yielding loss of synchronism of the generator. In other words, we will consider the system to be insecure if the CCT is smaller than 155ms, secure otherwise.

In order to illustrate automatic learning algorithms, we will generate by numerical simulation a data base of examples of diverse pre-fault operating conditions of the OMIB system, determining for each one its CCT and classifying it accordingly into the secure or insecure class. Note that only a part of this data base will be used as a learning set to train the automatic learning methods, the remaining part will be used as a test set to evaluate their reliability.

In the above OMIB system the following parameters may influence the security of the system : the amount of active and reactive power of the generator (denoted by Pu and Qu), the amount of load nearby the generator (Pl), voltage magnitudes at the load bus and at the infinite bus (Vl and Vinf), and also the short-circuit reactance Xinf, representing the effect of variable topology in the large system represented by the infinite bus. However, for the sake of simplicity we assume that only the active and reactive power of the generator are variable, while keeping the voltages Vl and Vinf constant (and equal to 400kV).[1]

We have generated a data base of 5000 such states, by (randomly) sampling Pu and Qu uniformly in the ranges indicated at Fig. 1.1, computing the CCT of each case by dichotomic search using time-domain simulation, then classifying the states as secure if their CCT is larger than 155ms, insecure otherwise. Table 1.1 lists a sample of states thus generated together with the corresponding CCT values; Figure 1.2 shows

Table 1.1. Sample of OMIB operating states

State Nb	Pu (MW)	Qu (MVAr)	Vl (p.u.)	Pl (MW)	Vinf (p.u.)	Xinf (Ω)	CCT (s)
1	876.0	-193.7	1.05	-100	1.05	60	0.236
2	1110.9	-423.2	1.05	-100	1.05	60	0.112
3	980.1	79.7	1.05	-100	1.05	60	0.210
4	974.1	217.1	1.05	-100	1.05	60	0.224
5	927.2	-618.5	1.05	-100	1.05	60	0.158
...	...	...	...	...	...	...	...
2276	1090.4	-31.3	1.05	-100	1.05	60	0.157
...	...	...	...	...	...	...	...
4984	1090.2	-20.0	1.05	-100	1.05	60	0.158
...	...	...	...	...	...	...	...

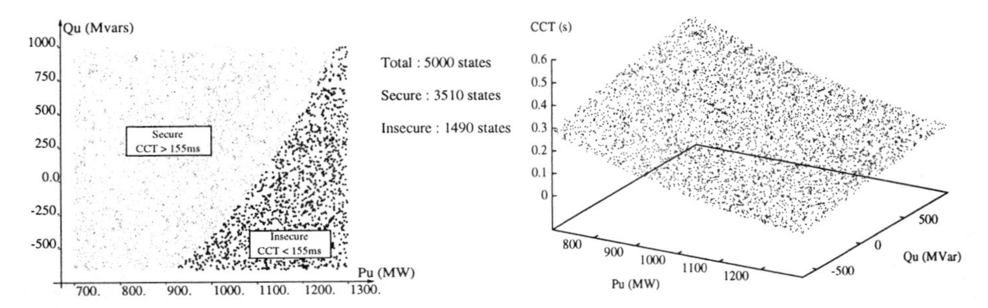

Figure 1.2. Data base of 5000 random states of the OMIB system

the scatter plots, illustrating how Pu and Qu act upon security (security class in the left hand part, CCT in the right hand part).

Learning set (LS). We use a random subsample of 3000 (Pu, Qu) states together with the output security information (class or CCT, as appropriate).

Test set (TS). We use the remaining 2000 states to evaluate the reliability of the different methods.

1.2.3 Symbolic knowledge via machine learning

Machine learning (ML) is a subfield of automatic learning concerned with the automatic design of rules similar to those used by human experts (e.g. if-then rules). We will describe only *Top down induction of decision trees* (TDIDT), which is one of the most successful classes of such methods [BFOS84, Qui93].

Decision trees. Before describing how TDIDT proceeds to build decision trees let us explain what a decision tree is and how it is used to classify a state. Figure 1.3 shows a hypothetical binary decision tree (DT) for our problem using the two attributes Pu

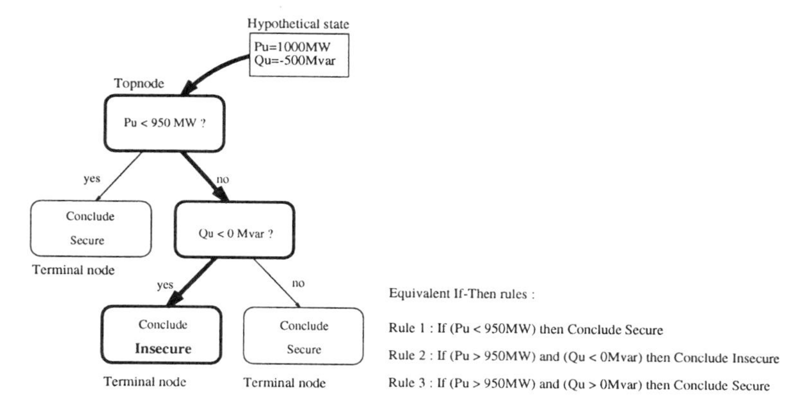

Figure 1.3. Hypothetical decision tree and equivalent if-then rules

and Qu. The bold arrows on the tree suggest how a hypothetical state (Pu = 1000 MW and Qu=-500 MVAr) traverses the tree in a top down fashion to reach a terminal node. One starts at the topnode and applies sequentially the dichotomous tests encountered to select the appropriate successor. When a terminal node is reached, the output information stored there is retrieved. Thus, for our hypothetical state the conclusion is "insecure". Note that the tree may be translated into an equivalent set of if-then rules, one for each terminal node. E.g. the tree in Fig. 1.3 translates into the rules indicated beneath it.

Decision tree growing. Now, let us illustrate on our example how the TDIDT method will extract from our learning set a set of classification rules in the form of a decision tree.

Figure 1.4 illustrates the successive node splitting procedure. The procedure is initialized by creating the topnode of the tree, which corresponds to the full LS as shown in Fig. 1.4a. Note that the relative size of the dark and light areas of the box used to represent the topnode corresponds to the proportion of insecure and secure states in the full learning set (909 insecure states vs 2091 secure states).

The method starts with a list of attributes (also called *candidate attributes*) in terms of which it will formulate the tree tests. In our example we use only two candidate attributes (Pu and Qu) since all other parameters are constant.

To develop the topnode each candidate attribute (here Pu and Qu) is considered in turn, in order to determine an appropriate threshold. To this end, the learning set is sorted by increasing order of the considered attribute values, then for each successive attribute value a dichotomic test is formulated and the method determines how well this test separates secure and insecure states, using an information theoretic score measure. The score measure is normalized, between 0 (no separation at all) and 1

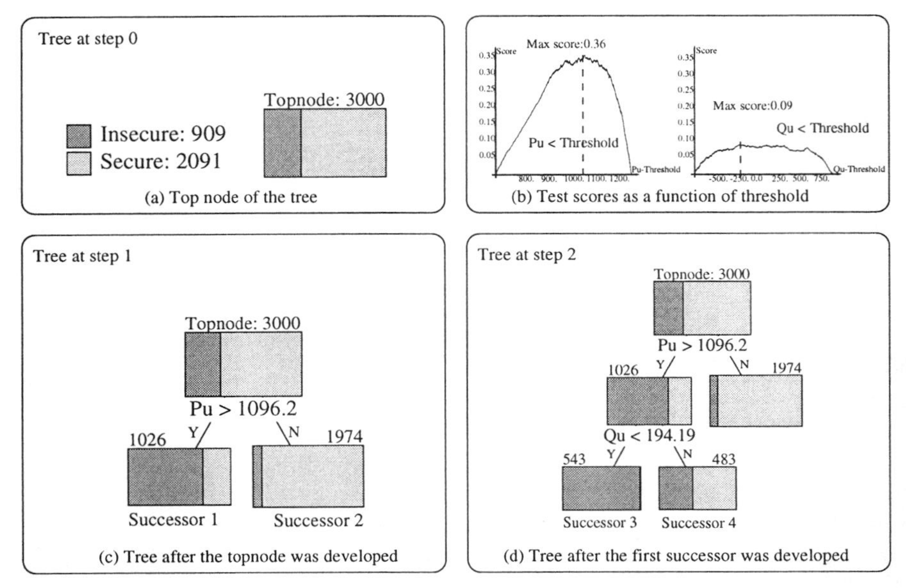

Figure 1.4. Three first steps of decision tree growing

(perfect separation). Figure 1.4b shows how the score varies in terms of the threshold both for Pu and Qu at the topnode. Thus, the optimal threshold for Pu is found to be 1096.2 MW (with a score of 0.36) and the optimal threshold for Qu is found to be -125MVAr (with a score of 0.09). Finally, the overall best test is identified at the topnode to be Pu>1096.2 MW.

Once the optimal test is found, the next step consists of creating two successor nodes corresponding to the two possible issues of the test; the learning set is then partitioned into corresponding subsets by applying the test to its states. The result of this step is represented at Fig. 1.4c. Note that the number on the top of each node represents the number of corresponding learning states : 3000 at the topnode, 1026 at the first successor and 1974 at the second successor. Note also that the first successor contains a strong majority of insecure states, while the second successor contains a very strong majority of secure states.

Stopping to split criterion. As is illustrated on Fig. 1.4d, the procedure continues recursively to split the recently created successors, gradually separating the secure and insecure states until a stop splitting criterion is met. The stop splitting criterion decides whether a node should indeed be further developed or not. There are two possible reasons to stop splitting a node, which yield two types of terminal nodes : *leaves* and *deadends*. A leaf is a node which corresponds to a sufficiently pure subset

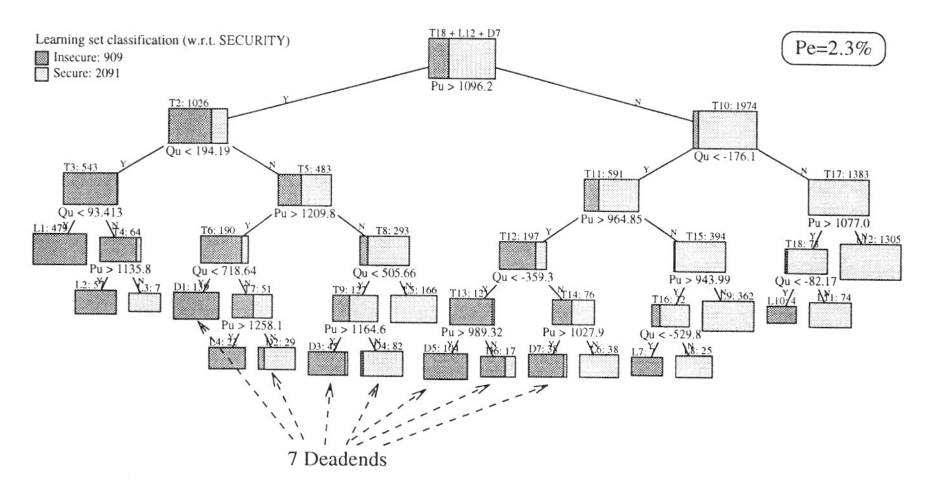

Figure 1.5. "Orthogonal" decision tree (end result)

(e.g. all states belong to the same class). A deadend is a node where there is not enough statistical support for choosing an appropriate test. Stop splitting at deadend nodes prevents the tree from over-fitting the learning set and hence allows the method to reach a good compromise between accuracy and simplicity.

The end result of this procedure is the tree shown at Fig. 1.5, which partitions the learning set into subregions defined by line segments orthogonal to the Pu or Qu axes; this "orthogonal" tree is composed of 18 test nodes, 12 leaves and 7 deadends.

Validation. Since the tree is grown to reach a good compromise between simplicity and separation of secure and insecure *learning* states it provides a kind of summary of the relationship observed in the learning set between Pu and Qu attributes and security class. But, how well does it generalize to unseen states ? To answer this question, we use the test set of 2000 states different from the learning states and compare the security class predicted by the tree with the one derived from the CCT computed by numerical simulation.

Thus each test state is directed towards a terminal node on the basis of its input attribute values (Pu and Qu) and applying sequentially the dichotomous tests encountered to select the appropriate successor. When a terminal node is reached, the output majority class of the corresponding learning subset stored there is retrieved and the test state is classified into this class. E.g. states reaching terminal nodes L1, L2, D1, L4, D3, D5, D6, D7, L7 and L10 are predicted to be insecure, while those reaching terminal nodes L3, D2, D4, L5, L6, L8, L9, L11 and L12 are predicted to be secure. Among the 2000 test states, this yields 1954 correct classifications, 15 insecure states declared erroneously secure, and 31 false alarms, i.e. an error rate Pe of 2.3%.

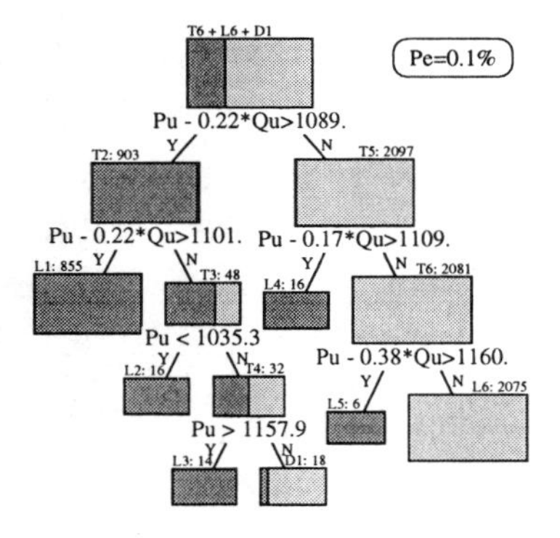

Figure 1.6. "Oblique" decision tree

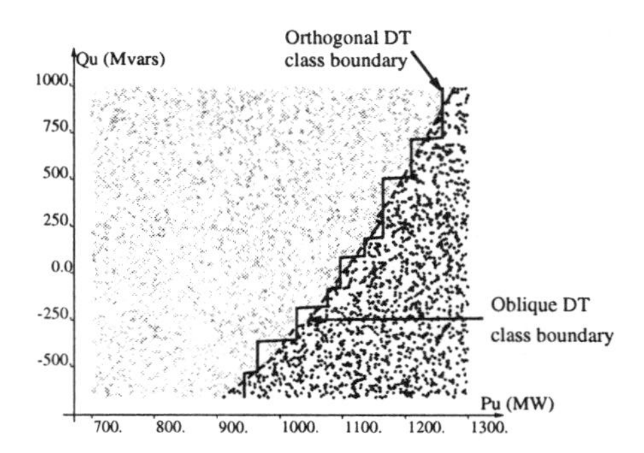

Figure 1.7. Classification boundaries of the DTs of Figs. 1.5 and 1.6

Refinements. There are many refinements of the TDIDT method of interest in the context of security assessment. First of all, decision trees may exploit easily discrete attributes (e.g. to represent power system topology) together with numerical ones. They may also be generalized to an arbitrary number of (security) classes and to tests with more than two outcomes.

Another interesting extension consists of using linear combinations instead of single attribute (orthogonal) splits, yielding so-called "oblique" decision trees. They are useful when there are strong interactions among different candidate attributes. For example, in our illustrative problem we could use linear combinations among Pu and Qu, which should provide a more efficient separation between secure and insecure states.

Figure 1.6 shows a tree obtained in this fashion. During tree building, we search for splits in the form of "Pu + Weight*Qu<Threshold" instead of searching for single attribute splits (in the form of "Pu<Threshold" and "Qu<Threshold"). The optimal splitting procedure is modified in order to determine automatically both an appropriate weight and the optimal threshold at each test node.

The fact that the resulting "oblique" tree is significantly simpler than the "orthogonal" one of Fig. 1.5 (only 6 test nodes, 6 leaves and 1 deadend) confirms our intuition. The tree is also much more reliable (no non-detections and only two false alarms among the 2000 test states, i.e. an error rate of 0.1%).

Figure 1.7 further illustrates the difference between the two classification boundaries induced by the two trees : a rather rough staircase approximation for the "orthogonal" tree vs a much smoother boundary for the "oblique" one.

The only price to pay for this improvement is an increase in CPU time at the tree growing stage, since searching for linear combinations is more intricate than searching for optimal thresholds. E.g. in our example it took 120 seconds[2] to grow the "oblique" tree and only 13 seconds to grow the "orthogonal" one.

In addition to "oblique" trees, other interesting extensions are *regression* trees which infer information about a numerical output variable, and *fuzzy* trees which use fuzzy logic instead of crisp logic to represent output information in a smooth fashion (see §5.5). Both approaches allow us to infer information about security margins, similarly to the techniques discussed below in §§1.2.4 and 1.2.5.

Salient features of decision trees. The main strength of decision trees is their interpretability. By merely looking at the test nodes of a tree one can easily sort out the most salient attributes (i.e. those which most strongly influence the output) and find out how they influence the output. Furthermore, at the tree growing stage the method provides a great deal of additional information, e.g. about scores of different candidate attributes, their correlations, and the overall information they provide to the tree (see Chapter 5).

Another very important asset is the ability of the method to identify the most relevant attributes for each problem. Our toy problem was too simple to illustrate this feature, but in large-scale applications less than twenty percent of the candidate attributes are typically selected while growing a tree (see Chapter 10).

The last characteristic of decision trees is computational efficiency : tree growing computational complexity is practically linear in the number of candidate attributes and in the number of learning states, allowing one to tackle easily problems with a few hundred candidate attributes and a few thousand learning states. The use of a tree as a classification algorithm is ultrafast, compatible with any real-time constraints.

Computational efficiency together with interpretability enable the method to be used in an interactive trial and error fashion, so as to discover interesting information contained in a data base and thereby gain physical insight into a problem. Below, we describe methods which are essentially complementary to decision trees and may be combined with them in hybrid approaches.

1.2.4 Smooth nonlinear approximations via artificial neural networks

The field of artificial neural networks (ANNs) has grown to an important and productive research field. Below, we restrict ourselves to multilayer perceptrons; other types of neural networks are discussed further in Chapter 4.

We mention also here that in addition to neural networks there exists a large diversity of modern statistical regression techniques, such as projection pursuit regression [FS81] and multivariate adaptive regression splines (MARS) [Fri91]. These methods are very similar to multilayer perceptrons in their approach and sometimes even more powerful (see Chapter 3).

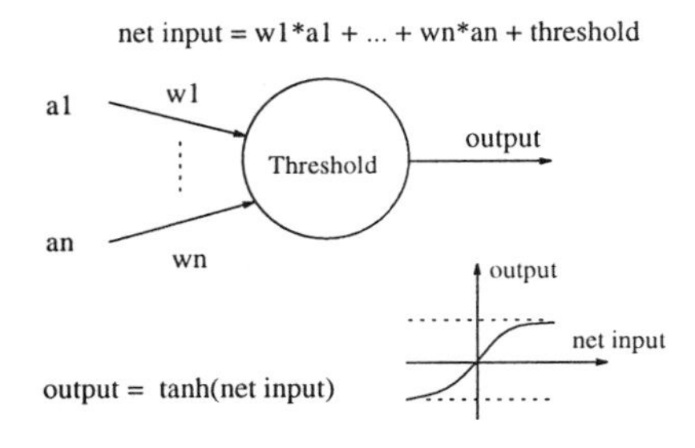

Figure 1.8. Perceptron (neuron)

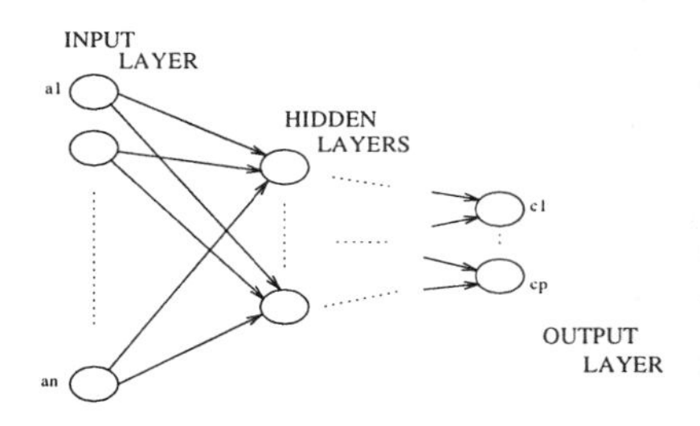

Figure 1.9. Multilayer perceptron

Multilayer perceptrons. Artificial neural networks' development started several decades ago with the work on perceptrons. Figure 1.8 illustrates the perceptron; this is basically a simple linear threshold unit, thus able to represent only linear boundaries in the attribute space. Its limited representation capabilities have motivated the consideration of more complex ANNs, composed of multiple interconnected layers of perceptrons, multilayer perceptrons (MLPs) for short. These latter are able to represent nonlinear input/output functions in a very flexible way.

Figure 1.9 illustrates a typical multilayer perceptron. Each neuron is a perceptron : input layer neurons are fed with the input attributes; hidden and output layer neurons receive linear combinations of outputs from neurons in the preceding layers.

Learning. In the context of multilayer perceptrons, the learning stage consists of determining an appropriate structure of the MLP and of identifying appropriate values of the different parameters (weights and thresholds).

Choosing the structure consists of defining the number of neurons, the topology of their interconnection, and the type of activation functions they use. Usually, it is determined by a trial and error procedure. However, nowadays there exist various algorithms to determine the structure automatically (see §§3.3 and 4.3.7).

The parameter identification task amounts to a complex nonlinear numerical optimization problem, which may be solved by various techniques. Historically, the first method which was proposed was the so called "back-propagation" algorithm, which is equivalent to a fixed step gradient descent technique. It is interesting from a biological point of view, but rather inefficient from a computational point of view. Nowadays, one uses generally second order quasi-Newton methods.

Parameter identification. In our illustrative problem MLPs can be exploited interestingly to approximate the CCT as a closed form function of Pu and Qu. Generally, for the approximation of a continuous function, MLPs with a single hidden layer may provide good approximators (see §4.3.6). Thus, let us try to approximate the CCT with

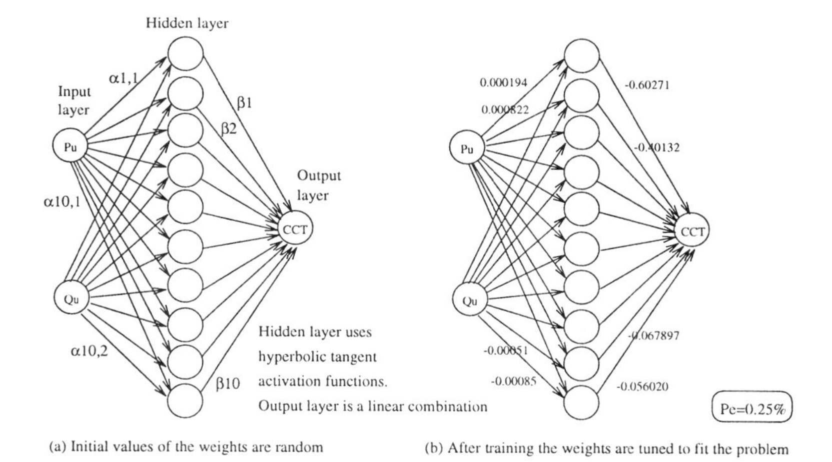

(a) Initial values of the weights are random (b) After training the weights are tuned to fit the problem

Figure 1.10. Single hidden layer perceptron

such a structure. The learning set used will be the same 3000 input states that we used for building the decision trees, but we associate as output information to each state its CCT, rather than the security class.

Figure 1.10 graphically sketches the MLP that we have used, containing 10 hidden neurons. Each hidden neuron i has an input/output relationship of the form

$$\text{Output}_i(\text{state}) = \tanh(\alpha_{i,1} Pu(\text{state}) + \alpha_{i,2} Qu(\text{state}) + \theta_i), \tag{1.1}$$

where $\alpha_{i,1}$ (resp. $\alpha_{i,2}$) is the connection weight between the neuron and the Pu (resp. Qu) input, and θ_i its threshold.

The output of the MLP is a linear combination of the preceding functions, i.e.

$$\text{CCT}_{\text{MLP}}(\text{state}) = \sum_{i=1\ldots10} \beta_i \tanh(\alpha_{i,1} Pu(\text{state}) + \alpha_{i,2} Qu(\text{state}) + \theta_i), \tag{1.2}$$

where β_i represents the contribution of neuron i in the overall output.

The parameter identification thus aims at choosing appropriate values of the 40 parameters $(\alpha_{i,j}, \theta_i, \beta_i)$, in order to fit for each learning state the MLP output to the CCT value determined by numerical simulation. The fitting criterion we use is the mean square error (SE)

$$SE = N^{-1} \sum_{\text{state} \in \boldsymbol{LS}} |\text{CCT}(\text{state}) - \text{CCT}_{\text{MLP}}(\text{state})|^2, \tag{1.3}$$

which is a smooth, although complex and nonlinear function of the parameter values, which needs to be minimized.

Before starting the learning procedure the parameters are all initialized at random, then they are progressively adapted in order to minimize the SE. In our example we used the Broyden-Fletcher-Goldfarb-Shanno (BFGS) optimization method (see §4.3.5).

At initialization, the random initial parameter values of the MLP yielded a SE equal to 0.0689568683, corresponding to a root mean square (RMS) appromixation error of 0.2626 seconds, which is three times larger than the standard deviation of CCT values in the learning set. In short, the initial output values are random.

After 46 iterations the algorithm stops at a local minimum, having reduced the SE to 0.0000003136, which corresponds to a very small RMS appromixation error of 0.00056 seconds. The mean absolute error in the learning set is equal to 0.4 milliseconds which is actually smaller than the accuracy of the CCT values determined by numerical simulation. Thus, we deem that we are close to the global minimum.

All in all, the parameter adaptation process took 730 CPU seconds on a Sparc 10 SUN workstation.

The resulting closed form approximation of the MLP input/output function corresponding to the final parameter values is as follows

$$\begin{aligned} \mathrm{CCT}_{\mathrm{MLP}} = \; & -0.602710 \tanh(0.000194Pu - 0.00034Qu - 0.93219) \\ & -0.401320 \tanh(0.000822Pu - 0.00020Qu - 0.76681) \\ & +0.318249 \tanh(0.000239Pu - 0.00050Qu - 0.29351) \\ & -0.287230 \tanh(0.002004Pu - 0.00034Qu - 1.20080) \\ & +0.184522 \tanh(0.000131Pu - 0.00057Qu - 0.03152) \\ & +0.177701 \tanh(0.001799Pu - 0.00011Qu - 2.08190) \\ & -0.150720 \tanh(0.001530Pu - 0.00056Qu - 1.68040) \\ & +0.142678 \tanh(0.002152Pu - 0.00046Qu - 1.72280) \\ & -0.067897 \tanh(0.001910Pu - 0.00051Qu - 1.71343) \\ & -0.056020 \tanh(0.000202Pu - 0.00085Qu - 0.39876) \end{aligned} \tag{1.4}$$

Validation. In order to evaluate the reliability of this approximation, we have used the MLP to predict the CCTs of the 2000 test states. Figure 1.11 shows the distribution of errors; it is clear that in this simple example the MLP approximates the CCT with very high accuracy. Furthermore, we observe that the mean absolute error in the test set (0.4ms) is equal to the mean absolute error in the learning set. Thus, the MLP generalizes very well to unseen states.

Consequently, the MLP can be used in order to classify states with respect to a threshold. For example, with respect to the threshold of 155ms used in the decision trees, it classifies 5 insecure states as secure (their CCT is however very close to the threshold of 155ms) and makes no false alarms, i.e. its error rate is of 0.25%.

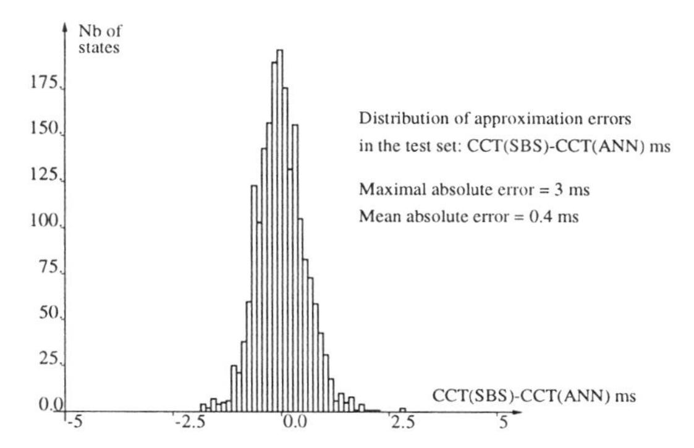

Figure 1.11. Distribution of MLP approximation errors (test set)

Note that since the MLP provides a very accurate closed form approximation of the CCT, its sensitivities with respect to Pu and Qu may be computed analytically. These could in turn be used to find out preventive control actions to increase the value of the CCT whenever it is found to be too small.

Refinements. Although in many problems a single hidden layer is sufficient, it is straightforward to generalize the MLP by adding any number of further layers. It is also possible to use other activation functions than the hyperbolic tangent, e.g. Gaussian or trigonometric functions.

Another extension consists of growing the neural network by progressively adding neurons and/or layers. Pruning techniques, with fancy names like "optimal brain dammage" and "optimal brain surgeon", were also designed to reduce over-fitting by removing useless connections and neurons. There are even techniques (e.g. the projection pursuit regression method already mentioned) able to adapt automatically the shape of the activation functions to the problem features. In Chapter 6, we will describe another technique to determine the network structure, which uses a hybrid approach combining decision trees with multilayer perceptrons.

Finally, let us mention that the MLP learning algorithm may be used in an adaptive on-line scheme, so as to adapt parameters whenever new learning states become available.

Salient features of MLPs. Notice that the very high accuracy obtained in our OMIB example is due to the fact that we used a very large learning set in order to approximate a rather smooth function of only two input parameters, and that we have used a moderate number of parameters in the MLP. Further, our OMIB data base is free of noise.

However, in many practical large scale applications the conditions are generally less favorable and it is not possible to reach this level of accuracy. Nevertheless, MLPs are often among the most reliable automatic learning methods.

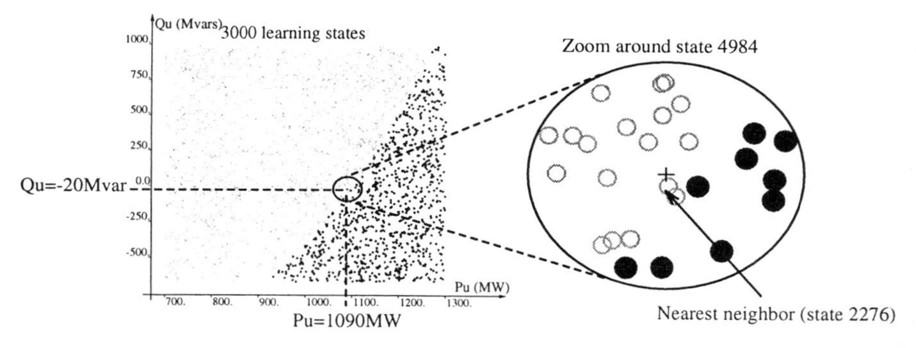

Figure 1.12. 3000 learning states and nearest neighbors of state 4984

Thus, the main characteristic of MLPs is flexibility in approximating nonlinear functions in multi-dimensional spaces.

This flexibility is however obtained at the expense of high computational burden. In real-life problems, when the number of inputs and hidden neurons is large, training times are typically of several hours to several days. At the same time, it becomes rather difficult to appraise and interpret the type of input/output relationship represented by such an MLP, which thus behaves like a black box.

1.2.5 Memory based reasoning via statistical pattern recognition

Decision trees and multilayer perceptrons essentially compress detailed information contained in their learning set into general, more or less global models (rules or real-valued functions).

Additional information may however be provided in a case by case fashion, by matching an unseen situation with similar situations found in the data base. This may be achieved by defining generalized distances so as to evaluate similarities among power system situations, together with appropriate fast data base search algorithms. Let us briefly illustrate the so-called "K nearest neighbors" (KNN) method.

KNN consists of classifying a state into the majority class among its K nearest neighbors in the learning set. In its most simple version, the learning stage of the KNN method merely consists of storing the learning states in a table. The actual work (computing the distances and sorting out the K nearest neighbors) is done when the method is used to make predictions for unseen states.

For example, in our illustration let us consider the state no 4984 of our data base (a test state). Its values of Pu and Qu are respectively of 1090 MW and -20 MVAr. Figure 1.12 shows in its left hand part the location of this state in the attribute space together with the learning states. In the right hand part we have zoomed on the nearest neighbors of the state. Note that the points on the borderline of the zoom region are equidistant (Euclidean distance) to the test state.[3] One may identify on Fig. 1.12 the

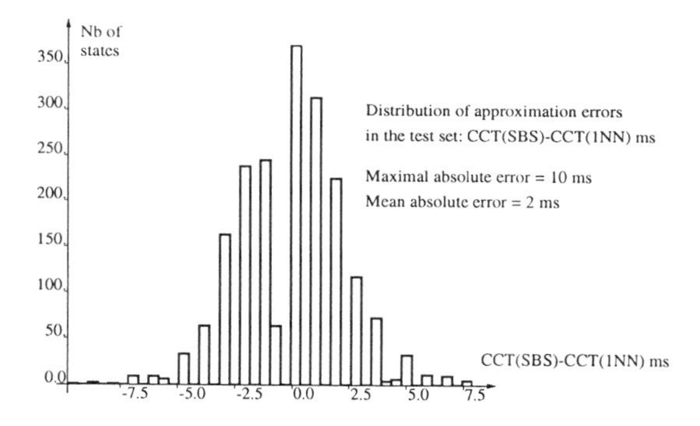

Figure 1.13. Distribution of 1NN approximation errors (test set)

nearest neighbor, i.e. the learning state closest to the test state (state no 2276 : Pu=1090 MW, Qu=−31 MVAr, and CCT=0.157s). Thus, according to the 1 nearest neighbor (1NN) rule, the CCT of the test state will be approximated to 0.157s and it would be classified into the secure class. Note that its actual CCT is equal to 0.158s; hence the state is correctly classified, in spite of being very close to the security boundary.

Validation. Repeating this procedure for all 2000 test states yields an error rate of 0.9%. Figure 1.13 shows the distribution of CCT approximation errors. Comparing with Fig. 1.11, we notice that the 1NN approximation is slightly less accurate than the MLP approximation. On the other hand, the 1NN provides additional information to that of the MLP and the DT : the distance to the nearest neighbors, attribute values of the nearest neighbors, and more generally any type of information attached to the nearest neighbors, like, for example, optimal preventive or emergency control strategies.

Refinements. The basic refinement consists of using K neighbors instead of a single one, K being determined on the basis of the learning set to increase reliabilty. Then, since the nearest neighbor rule is quite sensitive to the distance chosen, in many practical problems it is necessary to down weight less relevant attributes and enhance more relevant ones. Thus, distance learning algorithms have been devised so as to choose automatically the weights on the basis of a learning set. A further refinement consists of using different distance definitions in different regions of the attribute space.

In Chapter 10, we will illustrate some of these features on a real large-scale problem.

Salient features of KNN. The main characteristics of this method are high simplicity but also sensitivity to the type of distances used. In particular, to be practical, ad hoc algorithms must be used to choose the distances on the basis of the learning set. One such method will be briefly described in Chapter 6.

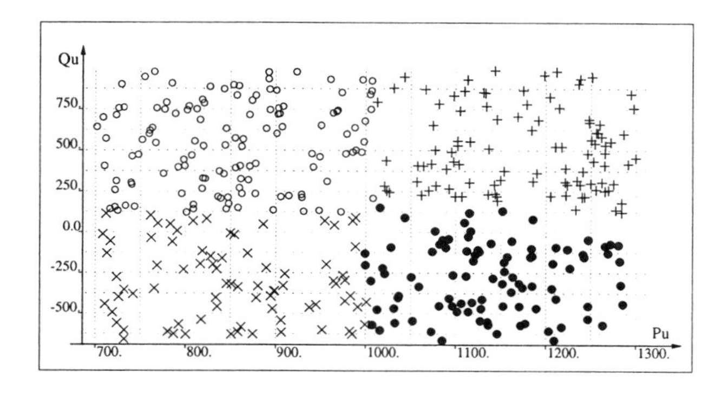

Figure 1.14. OMIB system Pu-Qu space clustering illustration

The fact that the KNN approach is quite similar to human reasoning (recalling similar situations seen in the past) makes it also interpretable by human operators.

1.2.6 Unsupervised learning

In contrast to supervised learning, where the objective is clearly defined in terms of modeling the underlying correlations between some input variables and some particular output variables, unsupervised learning methods are not oriented towards a particular prediction task. Rather, they try to find out by themselves the existing relationships among states characterized by a set of attributes.

Thus, one of the purposes of clustering is to identify homogeneous groups of similar states, in order to represent a large data base by a small number of representative *prototypes*. Graphically, two-dimensional scatter plots may be used as a tool in order to analyze the data and identify clusters. For example, Figure 1.14 shows the type of clustering obtained in the Pu-Qu space with the K-means algorithm (see §3.6.2), using 4 clusters.

Another application of the same techniques is to identify correlations (and redundancies) among the different attributes used to characterize states. In the context of power system security both applications may be useful as complementary data analysis and preprocessing tools.

Unsupervised learning algorithms have been proposed under the three umbrellas given above to classify classification methods, termed *cluster analysis* in the statistics literature, *conceptual clustering* in the machine learning community, and *self-organizing maps or vector quantization* in the neural net community [Koh90].

Unsupervised learning methods become really useful only in the context of large scale data bases, containing several thousand states described by many attributes. We will have to wait until the presentation of a real-life power system security problem in Chapter 10 to provide an interesting illustration.

Notes

1. Admittedly, to be more realistic we could introduce the effect of the other mentioned parameters, Pl, Vl, Xinf and Vinf by making them vary according to appropriate random distributions.

2. CPU times on a SUN Sparc10 workstation.

3. The equidistant region is slightly oval due to the fact that we have normalized Pu and Qu by their standard deviation before computing the distance.

I AUTOMATIC LEARNING METHODS

2 AUTOMATIC LEARNING IS SEARCHING A MODEL SPACE

Automatic learning aims at extracting information in various forms from *data bases*. The three paradigms to automatic learning are (i) statistics (pattern recognition, regression, density estimation), (ii) machine learning (concept learning from examples, conceptual clustering), and (iii) artificial neural network based learning. Although many of the theoretical and practical problems studied in these three fields are similar, and have received similar solutions, the three research communities have been rather isolated in the past. Unification of these paradigms, both theoretical and practical, was initiated only in the late eighties, when machine learning researchers started adopting probabilistic approaches and statisticians became interested in the new developments in the field of artificial neural networks.

The main unifying concept in automatic learning is to view it as a *search process* in a space of candidate *models*[1] (or *hypotheses*). The search process aims at identifying (or constructing) a model of maximal *quality*, and is guided by the information contained in the *learning set* (a subset of the data base) and possibly some background knowledge about the practical problem domain. Thus, the characteristics of a given automatic learning method depend on : the structure of its hypothesis space, the quality measure

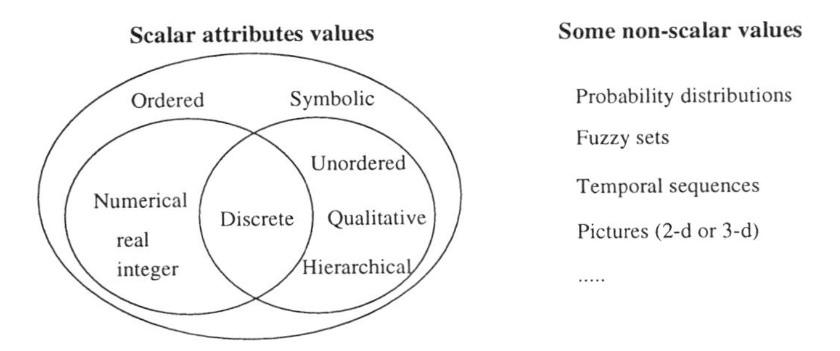

Figure 2.1. Types of attribute values

chosen to evaluate candidate hypotheses, and the search strategy used to explore the hypothesis space.

The purpose of this chapter is to describe these generic concepts which hold throughout the different paradigms and to introduce the notation used subsequently to describe various automatic learning algorithms. The treatment is formal, based on a probabilistic framework, but we will stay at a fairly intuitive level and illustrate the introduced concepts through power system examples.

2.1 KNOWLEDGE DISCOVERY IN DATA BASES

A *data base* (DB) as we view it, is a collection of *objects* described by a certain number of *attributes* providing some information. For example, in power system load forecasting, objects may represent time series of consumptions at different nodes and at different time instants; attributes would be successive values of the load, meteorological data and possibly information about the type of consumers to which the load refers. Thus attributes may have different types of possible values as suggested at Fig. 2.1.[2]

Knowledge discovery in data bases (KDD) is the non-trivial process of identifying valid, novel, potentially useful, and ultimately understandable patterns in data [FPSSU96]. Another frequently used term is *Data mining* (DM) which is a step in the KDD process consisting of using an automatic learning method on a well defined learning problem in order to build patterns (or models) in a particular family.

2.1.1 Different subtasks of data mining

In general, data mining comprises the following subtasks [MST94].

Representation consists of (i) choosing appropriate input attributes to represent the practical problem instances, (ii) defining the output information, and (iii) choosing a class of models suitable to represent input/output relations (e.g. decision trees or multilayer perceptrons).

Attribute selection aims at reducing the dimensionality of the input space by dismissing attributes which do not carry useful information to predict the considered output information.

Model selection (or learning per se) will typically identify in the predefined class of models the one which best fits the learning states. This generally requires choice of model structure and parameters, using an ad hoc search (optimization) technique adapted to the considered type of model.

Interpretation and validation are very important in order to understand the physical meaning of the synthesized model and to determine its range of validity. It consists of comparing the information which can be derived from the model with prior expertise and of testing it on a set of unseen test examples.

Model use consists of applying the model to predict outputs of new situations from the values assumed by the input parameters, and if necessary to "invert" the model in order to provide information on how to modify input parameters so as to achieve a given output.

Solving the *representation problem* is normally left to the engineer, although some methods are able to construct new attributes automatically. Thus choosing an appropriate set of attributes leads to an iterative process during the first trials of applying a learning algorithm to a new problem. Similarly, the choice of a suitable type of models is done in a trial and error fashion. Notice also that it may be necessary to preprocess the attribute values in order to apply a given learning algorithm, e.g. to scale attribute values, to code symbolic information numerically, or to estimate missing values.

The distinction between *attribute selection* and *model selection* is somewhat arbitrary. For example some of the methods (in particular decision tree induction) actually solve these two problems simultaneously.

From the *interpretation and validation* point of view, we have already seen that some methods provide rather black-box information, difficult to interpret, while some others provide more transparent explicit models, easier to compare with prior knowledge.

Finally, as far as the *use* of the model for fast decision making is concerned, we mention that speed variations of several orders of magnitude may exist between various techniques. This may reduce the usefulness of some methods in time-critical real-time applications.

Notice that in the data mining process, intensive use is made of various *graphical visualization* techniques. In our illustrations we will provide many graphical visualization examples, some being general tools (e.g. frequency histograms, scatter plots, dendrograms ...) some being specifically developed for power system applications (e.g. using one-line diagrams to represent zones or security scenarios).

2.1.2 *Different applications of automatic learning*

One may distinguish among the three following types of applications.

1. Descriptive : AL is used in order to describe the information contained in a DB. This concerns descriptive statistics, data summarization and graphical visualization as subtopics, which are important in practical applications.

2. Inductive : AL learning is used to infer general rules from a DB, which can then later be applied deductively to infer unknown information of new objects.

3. Transductive : AL learning is applied to a DB to directly infer an unknown information for a particular object (i.e. without formulating a general rule [Vap95]).

In this book we will mainly concentrate on the second type of applications which have led to a comprehensive set of general tools.

2.2 REPRESENTATION OF OBJECTS BY ATTRIBUTES

2.2.1 *Terminology and notation*

In our framework, we consider a DB as a (generally proper) subset of a *universe* $\boldsymbol{U}$ of all possible objects o. An attribute, denoted by $a(\cdot)$ is a function defined on $\boldsymbol{U}$; its value for an object $o \in \boldsymbol{U}$ is denoted by $a(o)$. Further, $a(\boldsymbol{X})$ denotes the set of all possible values that $a(\cdot)$ takes in $\boldsymbol{X} \subset \boldsymbol{U}$ and for any subset $\boldsymbol{V}$ of $a(\boldsymbol{U})$, we denote by $a^{-1}(\boldsymbol{V})$ the set of objects $\{o \in \boldsymbol{U} | a(o) \in \boldsymbol{V}\}$. In the sequel we will use upper-case boldface letters to represent sets of objects (e.g. $\boldsymbol{X}$ and $\neg \boldsymbol{X}$ its complement) or of attribute values.

A *data base* is thus a subset of objects described by a certain number of attributes providing information about each object.[3] For example, in a power system security data base, objects can correspond to simulated security scenarios and attributes provide physical information on power system states, or about the dynamic behavior of some devices in each scenario.

In a data mining application we will denote by *candidate attributes* the subset of attributes which are being used as input variables for learning, and by *test attributes* or *selected attributes* those which are eventually used in the learned rule. We will denote by n the number of such candidate attributes.

Attributes are in principle very general functions. Most of the time we will use scalar (numerical or symbolic) attributes, but occasionally more complex non-scalar data structures can also be considered as attributes. For example, in some practical applications most of the attribute values will be temporal signals.

Note also that some of the attributes are explicitly stored in the data base (we will call them *ground or explicit attributes*), while others are computed on the fly (we will call them *function attributes*). Thus, in a data mining session the user may define *function attributes* in order to combine information from different other (ground or function) attributes. Function attributes may also be constructed automatically using learning algorithms. For example, a decision tree can be translated into a classification function (a symbolic attribute) defined on $\boldsymbol{U}$, computed from its test attributes. Function attributes, whether user-defined or automatically induced, may of course be exploited as candidate attributes in later data mining sessions.

2.2.2 Structure induced by an attribute

In AL we will generally exploit attributes in order to infer a structure on subsets of objects from the structure defined on the attribute values. Indeed, any attribute induces an equivalence relation on objects, two objects being "equivalent" if they share the "same" value for this attribute.

In particular, a qualitative attribute $a(\cdot)$ is defined as an attribute with a finite *unstructured* set of possible values; it thus defines a finite unstructured partition on $\boldsymbol{U}$

$$A = \{\boldsymbol{A}_1, \ldots, \boldsymbol{A}_q\} \tag{2.1}$$

where $\boldsymbol{A}_i = a^{-1}(v_i)$ and q denotes the number of different values of $a(\cdot)$.

On the other hand, partially (resp. totally) ordered attributes will induce a partial (resp. total) order on the equivalence classes. This additional structure may be exploited to generate meaningful candidate partitions on $\boldsymbol{U}$ exploited in AL.

For example, a numerical attribute may be used to sort a set of objects and define *binary tests* (partitions) in the form of

$$T_{a_{th}} = \{\boldsymbol{Y}, \boldsymbol{N}\}, \tag{2.2}$$

where $\boldsymbol{Y} \triangleq \{o \in \boldsymbol{U} | a(o) \leq a_{th}\}$ and $\boldsymbol{N} = \neg \boldsymbol{Y}$.

2.3 CLASSIFICATION PROBLEMS

A classification problem is a supervised learning problem where the output information is a discrete classification[4]. One refers also to these problems as *concept learning from examples* or *discrimination*.

2.3.1 Classes

We will denote by $\boldsymbol{C} \triangleq \{c_1, \ldots, c_m\}$ the set of possible, mutually exclusive classes of a classification problems, and by $c(o)$ the class of object o[5]. The number m of classes is in principle arbitrary but generally rather small. Since the classification of an object is unique, the following partition is defined on the universe by the classification function $c(\cdot)$

$$C = \{\boldsymbol{C}_1, \ldots, \boldsymbol{C}_m\} \ : \ \boldsymbol{C}_i \triangleq \{o \in \boldsymbol{U} | c(o) = c_i\}. \tag{2.3}$$

For example, in the context of security assessment, classes could represent different levels of security of a system; they are often defined indirectly via security margins and some thresholds. In this case, we will denote by $\tau_1 < \tau_2 < \ldots < \tau_{m-1}$ the $m-1$ corresponding threshold values.

2.3.2 Types of classification problems

A classification problem is said to be deterministic if to any object *representation* corresponds a single possible class. Thus, the attributes can in *principle* be used to determine the correct class of any object without any residual uncertainty.

In practice, there are various sources of uncertainty which will prevent most of the problems from being deterministic. For example, in large-scale power system security issues it is generally not desirable to take into account every possible effect on security, due to simplicity constraints. Another example of non-determinism which is often neglected, is due to the limited accuracy of a real-time information system which provides attribute values. In some other circumstances, it is simply not possible to obtain a good knowledge of the system state in order to predict its future evolution, e.g. due to modeling uncertainties.

In addition to the above distinction, the notion of classification may come with different meanings, according to the type of physical problems considered.

Diagnostic problems. Classes correspond to different types of populations, which are clearly defined a priori. For example, people suffering from a certain disease and those who do not, form two mutually exclusive classes. In diagnostic problems, the possible values assumed by attributes are a causal consequence of the class membership. For example, in diagnosing a disease a physician would look at attributes such as high body-temperature. Although in principle perfect classification is possible, actual performance is often limited by the information contained in such descriptive attributes.

Prediction problems. Classes correspond to some future state of a system, which is characterized by attributes obtained from its present state. Here, classes are a causal consequence of attributes, although one may distinguish between the deterministic case, where the class is a *function* of the attributes and situations where there exists some degree of non-determinism, either intrinsically or due to limited information contained in attributes.

2.3.3 Learning and test sets

We use lower-case boldface letters to denote vectors : thus, $\mathbf{x} = (x_1, \ldots, x_n)^T$ denotes an n-dimensional vector. Note that components of vectors may be numerical or not; if they are supposed to belong to a particular vector space, it will be mentioned explicitly.

An *example* is a classified vector of attribute values corresponding to an observed or simulated object. The *learning set* ($\boldsymbol{LS}$) is a subset composed of N examples corresponding to different objects drawn from the data base

$$\boldsymbol{LS} \triangleq \{(\boldsymbol{v}^1, c^1), (\boldsymbol{v}^2, c^2), \ldots, (\boldsymbol{v}^N, c^N)\}, \tag{2.4}$$

where the vector $\boldsymbol{v}^k = (v_1^k, v_2^k, \ldots, v_n^k)^T = \boldsymbol{a}(o_k)$ represents the attribute values of an object o_k and $c^k = c(o_k)$ its class. [6]

Similarly, the *test set* ($\boldsymbol{TS}$) is subset of M other examples, used in order to evaluate the quality of a decision rule derived on the learning set. We assume always that the learning and test sets are disjoint and drawn randomly from the data base.

2.3.4 Decision or classification rules

Hypothesis space. A *decision rule* $d(\cdot)$, or a *hypothesis* is a function assigning a value in $\boldsymbol{C}$ to any possible attribute vector in $\boldsymbol{a}(\boldsymbol{U})$:[7]

$$d(a_1(o), \ldots, a_n(o)) \text{ or simply } d(o) \;:\; \boldsymbol{U} \longmapsto \boldsymbol{C}. \tag{2.5}$$

A decision rule induces the partition $D = \{\boldsymbol{D}_1, \ldots, \boldsymbol{D}_m\}$ on $\boldsymbol{U}$, defined by

$$\boldsymbol{D}_i \triangleq d^{-1}(c_i) = \{o \in \boldsymbol{U} \mid d(o) = c_i\} \quad (i = 1, \ldots, m). \tag{2.6}$$

The *hypothesis space* $\mathcal{D}$ of an AL algorithm is the set of candidate decision rules in which it operates. Examples of hypothesis spaces are *the set of binary decision trees* or *the set of multilayer perceptrons* (see Chapters 4 and 5).

Rule quality. To evaluate decision rules, we suppose that a quality measure $Q(\cdot,\cdot)$ is defined, which assigns a real number $Q(d, \boldsymbol{X})$ to every decision rule[8] $d \in \mathcal{D}$, and every subset of $\boldsymbol{U}$:

$$Q(d, \boldsymbol{X}) \;:\; \mathcal{D} \times \boldsymbol{U}^{\boldsymbol{U}} \longmapsto]-\infty \ldots +\infty[, \tag{2.7}$$

where $\boldsymbol{U}^{\boldsymbol{U}}$ denotes the power set of $\boldsymbol{U}$, i.e. the set of all subsets of $\boldsymbol{U}$.

The higher the quality of a decision rule d in a subset $\boldsymbol{X}$, the more appropriate is this rule for solving the classification problem in this subset. To learn a decision rule implies a search of the hypothesis space $\mathcal{D}$, so as to find a decision rule maximizing the chosen quality measure, evaluated on the learning set. This quantity is referred to in the sequel as *apparent* quality $Q(d, \boldsymbol{LS})$.

Appropriate quality measures will be defined later on, but in general a quality measure will combine different elementary evaluation criteria, selected among the following ones.

Reliability. The reliability (or accuracy) of a decision rule d is a measure of the similarity of the partition D it induces on $\boldsymbol{X}$ and the goal classification partition C. Frequently, reliability is defined as the expected probability of misclassification, or more generally as the expected misclassification cost. We will use the notation $R(d)$ for reliability in $\boldsymbol{U}$.

Comprehensibility. If a decision rule has to be validated by a human expert or applied by an operator, then comprehensibility is often a key feature. The rather vague (and subjective) notion of comprehensibility is generally replaced in practice by a well defined (but also subjective) complexity measure. We will use the notation $C(d)$ to denote the hypothesis complexity. Examples of complexity measures are the number of nodes of a decision tree and the number of independent tunable parameters (weights and thresholds) of a multilayer perceptron.

Cost of implementation. The complexity of implementing a decision rule may be another important aspect. This may involve the computational complexity of the algorithm used to apply the rule; it may also take into account the complexity of obtaining the attribute values (e.g. measurement cost). We will not consider the implementation cost in this book.

If we look more globally at the KDD process, the following two aspects become equally important.

Cost of data base collection. In security assessment problems as in many other computer experiments, the time required to generate data bases and running simulations might become a practical limitation.

Complexity of learning. This corresponds to the computational requirements in terms of CPU time and memory, that must be fulfilled in order to learn a rule. In some real-time applications this may be a critical aspect and, as we will see, there may exist variations of several orders of magnitude among different methods.

2.4 REGRESSION PROBLEMS

A regression problem is a supervised learning problem, where the output information is a continuous numerical value (or a vector of such values), rather than a discrete class.

We will denote by $\boldsymbol{y}(\cdot) = (y_1(\cdot), \ldots, y_r(\cdot))$ an r-vector valued regression function, $\boldsymbol{y}(o)$ its value for a particular object and $\boldsymbol{y}(\boldsymbol{U})$ its range. Examples of regression variables in the context of security assessment could be various load power margins for voltage security, and various energy margins for transient stability.

We will denote by $\boldsymbol{r}(\cdot)$ a regression model, which is a function assigning a value in the output space $\boldsymbol{y}(\boldsymbol{U})$ to any possible attribute vector in $\boldsymbol{a}(\boldsymbol{U})$:

$$\boldsymbol{r}(a_1(o), \ldots, a_n(o)) \text{ or simply } \boldsymbol{r}(o) \quad : \quad \boldsymbol{U} \longmapsto \boldsymbol{y}(\boldsymbol{U}). \tag{2.8}$$

The term *regression technique* is used to denote AL methods inducing regression models. We will denote by $\mathcal{R}$ the hypothesis space of a regression method. Learning and test samples are defined in the same fashion as in classification problems. Regression model quality will use slightly different formulations (see §2.7.7).

An important practical difference between classification and regression is that in regression we essentially aim at modeling smooth input/output relationships whereas in classification we seek for a partition of the universe into a finite number of regions.

2.5 FROM CLASSIFICATION TO REGRESSION AND VICE-VERSA

In Chapters 3, 4 and 5 we describe various classification and regression techniques. It should be noted that both can be applied to classification problems and regression problems, by transforming these latter.

The standard way of encoding a classification variable $c(\cdot)$ into a regression variable consists of replacing it by a vector of class indicator variables

$$\boldsymbol{y}(o) = (y_1(o), \ldots, y_m(o)) \text{ where } y_i(o) \stackrel{\triangle}{=} \delta_{c(o),c_i}, \tag{2.9}$$

where $\delta_{c(o),c_i}$ is the Kronecker symbol, defined by

$$\delta_{c(o),c_i} = 1 \text{ if } c(o) = c_i, \text{ 0 otherwise.} \tag{2.10}$$

On the other hand, we already suggested that a continuous output variable can be discretized into a finite number of classes, by using a set of appropriately chosen thresholds.

2.6 UNSUPERVISED LEARNING PROBLEMS

In unsupervised learning we will distinguish between object clustering, which consists in defining automatically subsets of similar objects, and attribute correlation analysis, which consists in analyzing similarities among attributes. Below we introduce only the most usual similarity measures used in these applications.

2.6.1 Distances between objects in an attribute space

Most clustering algorithms require the definition of a similarity measure, or distance measure. The vector distance between two objects in the attribute space is defined by

$$\boldsymbol{\delta}(o_1, o_2) \stackrel{\triangle}{=} (\delta_{a_1}(a_1(o_1), a_1(o_2)), \ldots, \delta_{a_n}(a_n(o_1), a_n(o_2))), \tag{2.11}$$

where $\delta_a(a(o_1), a(o_2))$ denotes a predefined scalar distance between the values of an attribute.

The definition of the distance between two attribute values depends on the attribute type. In particular, for a numerical attribute the (weighted) difference $\delta_a(a(o_1), a(o_2)) =$

$w_a * (a(o_1) - a(o_2))$ is generally used, whereas for a symbolic attribute a difference table $\delta_a(v_i, v_j)$ is defined explicitly for each pair of possible values, such that $\delta_a(v_i, v_j) = -\delta_a(v_j, v_i)$ and $\delta(v, v) = 0$.

Given the definition of a distance between attribute values, for each attribute, the L^k-norm of the vector distance defines the scalar distance, or simply distance, between two objects

$$\Delta(o_1, o_2) \triangleq \sqrt[k]{\sum_{i \leq n} |\delta_{a_i}(a_i(o_1), a_i(o_2))|^k}, \tag{2.12}$$

where $k = 1$ for the *Manhattan* (or city-block) distance, $k = 2$ for the *Euclidean* distance and $k = \infty$ for the *maximum absolute deviation* distance.

Finally, the scalar distance between two sets of objects is accordingly defined by the lower bound "inf" of the distances between objects of the two sets

$$\Delta(\boldsymbol{X}_1, \boldsymbol{X}_2) \triangleq \inf\{\Delta(o_1, o_2) | o_1 \in \boldsymbol{X}_1 \wedge o_2 \in \boldsymbol{X}_2\}. \tag{2.13}$$

2.6.2 Attribute similarity

Similarity measures may also be defined between attributes, e.g. as generalized correlation coefficients.

Anticipating on the probability notation introduced below, we will define two such measures which depend on the subset of objects $\boldsymbol{X}$ used to compute them.

Correlation coefficient. It is used to measure the similarity between two *real* valued attributes, and defined by

$$|\rho(a_1, a_2)|(\boldsymbol{X}) \triangleq \frac{|E\{(a_1 - E\{a_1|\boldsymbol{X}\})(a_2 - E\{a_2|\boldsymbol{X}\})|\boldsymbol{X}\}|}{\sqrt{E\{(a_1 - E\{a_1|\boldsymbol{X}\})^2|\boldsymbol{X}\} E\{(a_2 - E\{a_2|\boldsymbol{X}\})^2|\boldsymbol{X}\}}}, \tag{2.14}$$

where $E\{\cdot|\boldsymbol{X}\}$ denotes the conditional expectation operator defined below in §2.7.2.

Normalized mutual information. It is used to compare qualitative attributes, in terms of the partitions they induce on $\boldsymbol{U}$, and is defined by

$$C_{A_1}^{A_2}(\boldsymbol{X}) \triangleq \frac{2I_{A_1}^{A_2}(\boldsymbol{X})}{H_{A_1}(\boldsymbol{X}) + H_{A_2}(\boldsymbol{X})}, \tag{2.15}$$

where $I_{A_1}^{A_2}(\boldsymbol{X})$ denotes the mutual information of the two attributes in $\boldsymbol{X}$, and $H_{A_1}(\boldsymbol{X})$ and $H_{A_2}(\boldsymbol{X})$ the entropies of the partitions they induce on $\boldsymbol{X}$ (see §2.7.4).

2.7 PROBABILITIES

Although learning may be defined in a purely deterministic fashion, as was the case with early machine learning and neural network formulations, it is now recognized that a probabilistic framework is practically unavoidable as soon as a certain level of generality is required. From a more "impressionist" point of view, by using the probabilistic framework, we adopt right from the beginning the idea that the quantitative evaluation of *uncertainties* is one of the first issues in the context of learning problems, which calls for an explicit probabilistic treatment. For a highly interesting philosophical treatment of what probability is all about, we very much recommend reading the book of E. T. Jaynes [Jay95].

In this section we introduce some notation and considerations related to a probabilistic interpretation of the learning problems.

2.7.1 *General probabilities*

For any $\boldsymbol{X} \subset \boldsymbol{U}$, we denote by $P(\boldsymbol{X})$ the prior probability of observing an object of $\boldsymbol{X}$, and $P(\boldsymbol{X}|\boldsymbol{Y})$ the conditional or posterior probability of an object to belong to $\boldsymbol{X}$ given the information that it belongs to $\boldsymbol{Y}$. Assuming that $P(\boldsymbol{Y}) > 0$, the conditional probability is defined by

$$P(\boldsymbol{X}|\boldsymbol{Y}) \triangleq \frac{P(\boldsymbol{X} \cap \boldsymbol{Y})}{P(\boldsymbol{Y})}. \tag{2.16}$$

To denote probability measures, we will use the notation dP or $p(a)da$, where $p(a)$ is the density function corresponding to a continuous probability measure.

2.7.2 *Random variables*

Roughly speaking, a random variable is a real-valued function defined on $\boldsymbol{U}$, e.g. a real-valued attribute or a regression variable, which maps probabilities defined for subsets of $\boldsymbol{U}$ to probabilities of subsets of the real line. A random variable may be continuous or not, according to the continuity of the probability measure induced on the real line.

The expectation of a random variable y is denoted by $E\{y\}$ and defined by

$$E\{y\} \triangleq \int_U y(o)dP. \tag{2.17}$$

Similarly, the conditional expectation given the information that $o \in \boldsymbol{X}$ is denoted by $E\{y|\boldsymbol{X}\}$, and defined by

$$E\{y|\boldsymbol{X}\} \triangleq \frac{\int_X y(o)dP}{P(\boldsymbol{X})}. \tag{2.18}$$

2.7.3 Classification

To simplify, we denote by $P^i(\boldsymbol{X})$ the conditional probability of $\boldsymbol{X}$ given that $c(o) = c_i$, i.e.

$$P^i(\boldsymbol{X}) \triangleq P(\boldsymbol{X}|\boldsymbol{C}_i). \tag{2.19}$$

To further simplify, we will denote by $\boldsymbol{p}(\boldsymbol{X}) = (p_1(\boldsymbol{X}), \ldots, p_m(\boldsymbol{X}))$ the vector of conditional class-probabilities, defined by

$$p_i(\boldsymbol{X}) \triangleq P(\boldsymbol{C}_i|\boldsymbol{X}), \tag{2.20}$$

and use $\boldsymbol{p} \triangleq (p_1, \ldots, p_m)$ to denote the vector of prior class probabilities, $p_i \triangleq p_i(\boldsymbol{U})$.

2.7.4 Entropies

Let us introduce some frequently used notions from information theory.[9]

The *entropy* associated to a partition of $P = \{\boldsymbol{P}_1, \ldots, \boldsymbol{P}_p\}$ of $\boldsymbol{U}$ is defined on any subset $\boldsymbol{X}$ by

$$H_P(\boldsymbol{X}) \triangleq - \sum_{i=1,\ldots,p} P(\boldsymbol{P}_i|\boldsymbol{X}) \log P(\boldsymbol{P}_i|\boldsymbol{X}). \tag{2.21}$$

Entropy is maximal in the case of uniform probabilities

$$H_P(\boldsymbol{X}) \leq - \sum_{i=1,\ldots,p} \frac{1}{p} \log \frac{1}{p} = \log p, \tag{2.22}$$

and it is minimal in case of complete certainty ($\boldsymbol{P}_j = \boldsymbol{P}$ for some j, and $\boldsymbol{P}_i = \emptyset$ for $i \neq j$):

$$H_P(\boldsymbol{X}) \geq - \sum_{i=1,\ldots,p} \delta_{ij} \log \delta_{ij} = 0, \tag{2.23}$$

where δ_{ij} denotes the Kronecker symbol and the limit value $\lim_{x \to 0^+} x \log x = 0$ is assumed.

Thus, entropy measures the *uncertainty* related to a partition, i.e. how difficult it is to guess to which subset of the partition an object belongs.

The most relevant particular cases are the *classification entropy* of a subset $H_C(\boldsymbol{X})$, defined by

$$H_C(\boldsymbol{X}) \triangleq - \sum_{i=1,\ldots,m} P(\boldsymbol{C}_i|\boldsymbol{X}) \log P(\boldsymbol{C}_i|\boldsymbol{X}), \tag{2.24}$$

and the entropy $H_A(\boldsymbol{X})$ of a qualitative attribute, defined by

$$H_A(\boldsymbol{X}) \triangleq - \sum_{i=1\ldots q} P(\boldsymbol{A}_i|\boldsymbol{X}) \log P(\boldsymbol{A}_i|\boldsymbol{X}). \tag{2.25}$$

The *joint entropy* of two partitions $P = \{\boldsymbol{P}_1, \ldots, \boldsymbol{P}_p\}$ and $Q = \{\boldsymbol{Q}_1, \ldots, \boldsymbol{Q}_q\}$, is the entropy of the partition $\{\boldsymbol{P}_i \cap \boldsymbol{Q}_j | i \leq p, j \leq q\}$, denoted by $P \cap Q$:

$$H_{P \cap Q}(\boldsymbol{X}) = -\sum_{i,j} P(\boldsymbol{P}_i \cap \boldsymbol{Q}_j | \boldsymbol{X}) \log P(\boldsymbol{P}_i \cap \boldsymbol{Q}_j | \boldsymbol{X}). \tag{2.26}$$

Notice that

$$H_{P \cap Q}(\boldsymbol{X}) \leq H_P(\boldsymbol{X}) + H_Q(\boldsymbol{X}) \tag{2.27}$$

and equality holds only if the two partitions are independent in $\boldsymbol{X}$, i.e. if

$$P(\boldsymbol{P}_i \cap \boldsymbol{Q}_j | \boldsymbol{X}) = P(\boldsymbol{P}_i | \boldsymbol{X}) P(\boldsymbol{Q}_j | \boldsymbol{X}) \quad : \quad \forall i \leq p, j \leq q.$$

Furthermore,

$$H_{P \cap Q}(\boldsymbol{X}) \geq \max\{H_P(\boldsymbol{X}), H_Q(\boldsymbol{X})\}. \tag{2.28}$$

2.7.5 Mutual information and conditional entropies

The *mutual information* of two partitions is defined by

$$I_P^Q(\boldsymbol{X}) \triangleq H_P(\boldsymbol{X}) + H_Q(\boldsymbol{X}) - H_{P \cap Q}(\boldsymbol{X}). \tag{2.29}$$

It is symmetric by definition (i.e. $I_P^Q(\boldsymbol{X}) = I_Q^P(\boldsymbol{X})$), positive and equal to zero only in case of independence. In addition, it verifies the following inequalities

$$I_P^Q(\boldsymbol{X}) \quad \leq \quad \min\{H_P(\boldsymbol{X}), H_Q(\boldsymbol{X}), H_{P \cap Q}(\boldsymbol{X})\}. \tag{2.30}$$

Further, it is easy to show that if the two partitions are equivalent then $H_P(\boldsymbol{X}) = H_Q(\boldsymbol{X}) = H_{P \cap Q}(\boldsymbol{X}) = I_P^Q(\boldsymbol{X})$.

Thus the quantity

$$C_P^Q(\boldsymbol{X}) \triangleq \frac{2 I_P^Q(\boldsymbol{X})}{H_P(\boldsymbol{X}) + H_Q(\boldsymbol{X})} \tag{2.31}$$

is a *symmetric similarity measure* between two partitions. $C_P^Q(\boldsymbol{X}) = 0$ only if the partitions are independent and $C_P^Q(\boldsymbol{X}) = 1$ only if they are equivalent.

The mean *conditional* (or posterior) entropy of partition P given partition Q, is denoted by $H_{P|Q}(\boldsymbol{X})$ and defined by

$$H_{P|Q}(\boldsymbol{X}) = \sum_j P(\boldsymbol{Q}_j | \boldsymbol{X}) H_P(\boldsymbol{Q}_j \cap \boldsymbol{X}). \tag{2.32}$$

It is easy to show that $H_{P \cap Q}(\boldsymbol{X}) = H_P(\boldsymbol{X}) + H_{Q|P}(\boldsymbol{X}) = H_Q(\boldsymbol{X}) + H_{P|Q}(\boldsymbol{X})$, thus

$$I_P^Q(\boldsymbol{X}) = H_P(\boldsymbol{X}) - H_{P|Q}(\boldsymbol{X}) = H_Q(\boldsymbol{X}) - H_{Q|P}(\boldsymbol{X}). \tag{2.33}$$

Whenever there is no ambiguity on the set $\boldsymbol{X}$ on which entropies and information quantities are computed we will use the simpler notation H_P instead of $H_P(\boldsymbol{X})$.

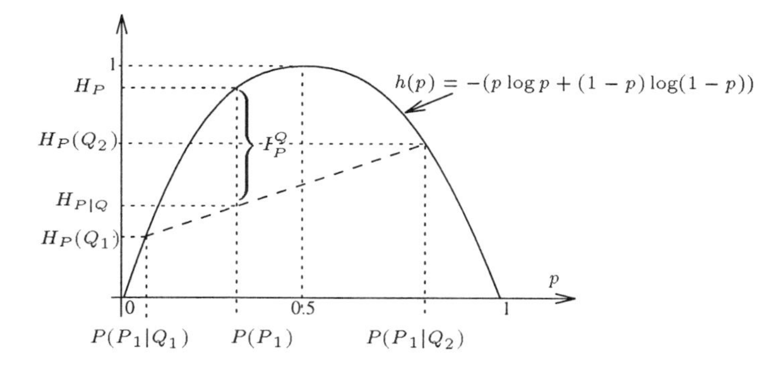

Figure 2.2. Geometric interpretation of information quantity

2.7.6 Geometric interpretation

Let us consider in Fig. 2.2 the particular case where we use information quantity to evaluate the "correlation" between two dichotomic (binary) partitions, say $P = \{\boldsymbol{P}_1, \boldsymbol{P}_2\}$ and $Q = \{\boldsymbol{Q}_1, \boldsymbol{Q}_2\}$. For example P could denote a two-class classification C and Q a dichotomic test T at a decision tree node (a "question" which partitions the corresponding subset of $\boldsymbol{U}$ into two subsets).

For the sake of simplicity, let us suppose that all entropies are computed over the whole universe ($\boldsymbol{X} = \boldsymbol{U}$). Thus prior (resp. conditional) classification entropies are obtained by the curve $h(p)$ of Fig. 2.2 replacing p by $P(\boldsymbol{P}_1)$ (resp. $P(\boldsymbol{P}_1|\boldsymbol{Q}_j), j = 1, 2$) :

$$H_P = h(P(\boldsymbol{P}_1)) \tag{2.34}$$

$$H_P(\boldsymbol{Q}_j) = h(P(\boldsymbol{P}_1|\boldsymbol{Q}_j)). \tag{2.35}$$

Moreover, the following identity holds for probabilities

$$P(\boldsymbol{P}_1) = P(\boldsymbol{Q}_1)P(\boldsymbol{P}_1|\boldsymbol{Q}_1) + P(\boldsymbol{Q}_2)P(\boldsymbol{P}_1|\boldsymbol{Q}_2) \tag{2.36}$$

and a similar one for the mean conditional entropy

$$H_{P|Q} = P(\boldsymbol{Q}_1)H_P(\boldsymbol{Q}_1) + P(\boldsymbol{Q}_2)H_P(\boldsymbol{Q}_2). \tag{2.37}$$

Thus on Fig. 2.2, $H_{P|Q}$ lies on the intersection of the vertical line at $P(\boldsymbol{P}_1)$ and the segment joining $(P(\boldsymbol{P}_1|\boldsymbol{Q}_1), H_P(\boldsymbol{Q}_1))$ to $(P(\boldsymbol{P}_1|\boldsymbol{Q}_2), H_P(\boldsymbol{Q}_2))$. Thus, I_P^Q is the vertical distance from this point to the $h(p)$ curve.

From this graphical interpretation we may induce the various properties given above : $0 \leq I_P^Q \leq H_P$ (the curve $h(p)$ is convex); $I_P^Q = H_P$ if and only if $P(\boldsymbol{P}_1|\boldsymbol{Q}_1) = 0$ and $P(\boldsymbol{P}_1|\boldsymbol{Q}_2) = 1$ or $P(\boldsymbol{P}_1|\boldsymbol{Q}_1) = 1$; $P(\boldsymbol{P}_1|\boldsymbol{Q}_2) = 0$; $I_P^Q = 0$ if and only if $P(\boldsymbol{P}_1|\boldsymbol{Q}_1) = P(\boldsymbol{P}_1|\boldsymbol{Q}_2) = P(\boldsymbol{P}_1)$.

2.7.7 *Reliabilities*

Decision rules. The evaluation of the reliability of a decision rule may rely on a loss matrix which allows one to weight different types of errors in a different way. Thus, given an $m \times m$ loss matrix $\boldsymbol{L}$, whose element L_{ij} defines the loss (or risk) corresponding to the decision c_j when the true class is c_i, the mean expected loss $L(d)$ of a decision rule is defined by

$$L(d) \triangleq \sum_{i=1}^{m} p_i \left[\sum_{j=1}^{m} L_{ij} * P^i(\boldsymbol{D}_j) \right]. \tag{2.38}$$

In the case of uniform misclassification cost, $L_{ij} = 1 - \delta_{ij}$, $L(d)$ reduces to the expected probability of misclassification $P_e(d)$, or to the complement of the *reliability*

$$R(d) = 1 - P_e(d). \tag{2.39}$$

Another evaluation of the reliability of a decision rule is based on the entropy concept, in terms of the mean information provided by a decision rule on the classification, i.e. I_C^D. In Chapter 5 we will use the *relative information* of a decision rule, defined by

$$RI_C^D \triangleq \frac{I_C^D}{H_C}. \tag{2.40}$$

Regression models. To evaluate reliability of regression models, we will generally use the least squares criterion,

$$SE(\boldsymbol{r}) = E\{||\boldsymbol{r} - \boldsymbol{y}||_2^2\}, \tag{2.41}$$

where SE stands for *square error* and $||\cdot||_2$ denotes the Euclidean norm. The smaller the square error, the more reliable the regression model is.

Another, frequently used reliability measure for regression problems is the expected *absolute error*

$$AE(\boldsymbol{r}) = E\{||\boldsymbol{r} - \boldsymbol{y}||_1\}, \tag{2.42}$$

where $||\cdot||_1$ denotes the L^1 norm.

Residual uncertainty and Bayes rules. For any classification (resp. regression) function defined on a universe $\boldsymbol{U}$, and a given choice of object representation in terms of attributes, there exists a theoretical upper bound on classification or regression performance. This theoretical upper bound can be derived from the knowledge of conditional probabilities of output values given attribute values (see for example [Cho91] for a complete treatment of this question). We will use the term *Bayes rule* to denote the corresponding model $\boldsymbol{d}_B(\boldsymbol{a})$ or $\boldsymbol{r}_B(\boldsymbol{a})$, and we will use the term *residual uncertainty* to denote its overall expected loss. Clearly, it depends on the probability assignment on subsets of $\boldsymbol{U}$.

Hence, if the sole objective is accuracy, the goal of automatic learning will be to guess the best possible approximation to the Bayes rule, from the limited information provided in the learning set and within the constraints imposed by the hypothesis space adopted.

2.7.8 Standard sample based estimates

We assume that the data base and its learning and test subsets are statistical independent samples drawn from the same probability distribution defined on $\boldsymbol{U}$. We assume also that their classification is a priori given and correct, as well as their attribute values. Thus, probabilities can be estimated from such samples.

However, it is important to realize that as soon as a learning set has been used to derive a decision rule or a regression model, any related estimates based on the learning set may become very unreliable. In particular, *apparent* quality estimates are generally very strongly optimistically biased.

Unless other information is to be taken into account, prior probability estimates of subsets of $\boldsymbol{U}$ are given by relative frequencies of these subsets in the learning or test sets. These estimates are substituted within reliability and entropy functions, to obtain the corresponding test or learning set estimates. Whenever necessary, we use the "hat" notation to distinguish the latter estimates from their true values in $\boldsymbol{U}$.

Expectation operators are replaced by sample means, unless specified otherwise. We use the "bar" notation to denote the sample mean of a random variable

$$\bar{x}(\boldsymbol{S}) \triangleq \frac{\sum_{o \in \boldsymbol{S}} x(o)}{|\boldsymbol{S}|}, \tag{2.43}$$

where $|\cdot|$ denotes the number of objects in a set (cardinality).

2.7.9 Various estimates of reliabilities

Below we discuss briefly the various types of reliability estimates used in practice. For a deeper discussion of the pros and cons of these methods, we invite the interested reader to refer to the literature (e.g. [Tou74, DK82, WK91] and the references therein). All these estimation procedures may be applied to any kind of reliability or cost measure used both in classification and regression problems, with trivial adaptations. Below we merely describe the case of estimating classification error rates.

Resubstitution estimate. The resubstitution estimate R^{LS} of reliability (or apparent reliability) is obtained from the learning sample itself. Since most learning algorithms try to identify a rule of maximal (or high) apparent reliability, this estimate is generally strongly optimistically biased, and does not provide in most practical situations any valuable information about the ability of the rule to classify unseen situations. In particular, it is not true in general that minimizing R^{LS} will also lead to minimal expected accuracy (see the discussion on over-fitting in §2.8).

Test set estimate. The test set estimate R^{TS} is obtained by using an independent sample to assess reliability. The independent test sample states are supposed to be correctly classified by a benchmark method (generally the same method which is used to classify the learning set). Their class is merely compared with the class predicted by the classification rule. This estimate is generally unbiased and, similarly to the resubstitution error estimate, its computation is straightforward.

The test set error probability estimate has a binomial sampling distribution, and for large sample sizes it is close to Gaussian. Its standard deviation may be estimated by the following formula

$$\hat{\sigma}_{\hat{P}_e} \approx \sqrt{\frac{\hat{P}_e(1-\hat{P}_e)}{M}}, \tag{2.44}$$

where $\hat{P}_e$ denotes the test set error estimate and M the size of the test set.

In particular if M is sufficiently large, a 95% confidence interval may be derived for the true error rate [DK82]

$$P\left\{\hat{P}_e - 1.96\hat{\sigma}_{\hat{P}_e} < P_e < \hat{P}_e + 1.96\hat{\sigma}_{\hat{P}_e}\right\} \approx 0.95. \tag{2.45}$$

For example, for a test sample size of 2000 and an estimated error rate of 3.0%, this interval is equal to [2.25% ... 3.75%].

For regression problems square error and absolute error test set estimates may be used. Again, the law of large numbers ensures that for sufficiently large test sets these estimates will assume Gaussian distributions centered around the true expected values, and simple formulas may be derived to estimate the standard errors of these estimates [BFOS84] enabling one to derive confidence intervals. For example the test set estimate of the absolute error is obtained by

$$\overline{AE}(\boldsymbol{TS}) = \overline{||\boldsymbol{r} - \boldsymbol{y}||_1}(\boldsymbol{TS}) \tag{2.46}$$

and its standard error estimate is obtained by

$$\hat{\sigma}_{\overline{AE}}(\boldsymbol{TS}) = \sqrt{\frac{\overline{(||\boldsymbol{r} - \boldsymbol{y}||_1 - \overline{AE}(\boldsymbol{TS}))^2}(\boldsymbol{TS})}{|\boldsymbol{TS}|}} \tag{2.47}$$

V-fold cross-validation estimate. V-fold cross-validation methods aim at providing an unbiased error estimate when no independent test set is available. Similarly to the resubstitution estimate, it exploits the $\boldsymbol{LS}$ used to build a decision rule d.

It partitions the $\boldsymbol{LS}$ into V non-overlapping (randomly selected) test sets $\boldsymbol{TS}_i, i = 1 \ldots V$ (approximately of size $\frac{N}{V}$). Each $\boldsymbol{TS}_i$ is then classified using a rule d_i built on a learning set $\boldsymbol{LS}_i = \bigcup_{j \neq i} \boldsymbol{TS}_j$, comprising the $V - 1$ remaining test sets. This provides V unbiased estimates of the error rate of classification rules determined on the basis of V slightly smaller learning sets than d. If V is not too small (e.g. $V \geq 10$) and each rule is built with the method used to derive the original one, the average error rate of these rules will reflect closely the true error rate of the original rule d.

The main disadvantage of this method is its high computational cost since it requires the repetitive learning of V different classification rules which may become overwhelming in the case of computationally intensive learning methods.

If V is equal to the number N of learning states this method reduces to the well known leave-one-out method. The leave-one-out method consists in building N successive models on the basis of N learning sets obtained by leaving one of the states out, and by testing the resulting model on the state left out and counting the total number of errors [WK91].

Choosing between the test set estimate and the cross-validation method is mainly a question of amount of available data. A rule of thumb is that below say 500 to 1000 available samples, dividing them into a test set and a learning set would either produce a too small test set, and thus high test set error estimate variances, or a too small learning set. Thus, we should probably prefer say 10-fold cross-validation and if less than 200 samples are available we could use the leave-one-out method [WK91].

Evaluation vs comparison of hypotheses. The above methods may be used to evaluate different hypotheses on the basis of the information contained in a data base. Thus, these methods may also be used in order to compare different models produced by different automatic learning methods. For practical purposes, a good rule of thumb is to declare as insignificant, reliability estimate differences which are smaller than the standard deviation of these latter.

However, the comparison of different methods calls for hypothesis testing methods specially designed for this purpose. We invite the interested reader to refer to [Die96] for further information on this topic.

2.8 OVER-FITTING : A GENERIC PROBLEM IN AUTOMATIC LEARNING

Over-fitting is a generic problem encountered in automatic learning. It generally appears when the model extracted is too complex with respect to the information provided in the learning set.

The complexity of a model measures the number of "parameters" in the model which are identified during learning. For example, the complexity of a decision tree is proportional to its number of terminal nodes, the complexity of a multilayer perceptron

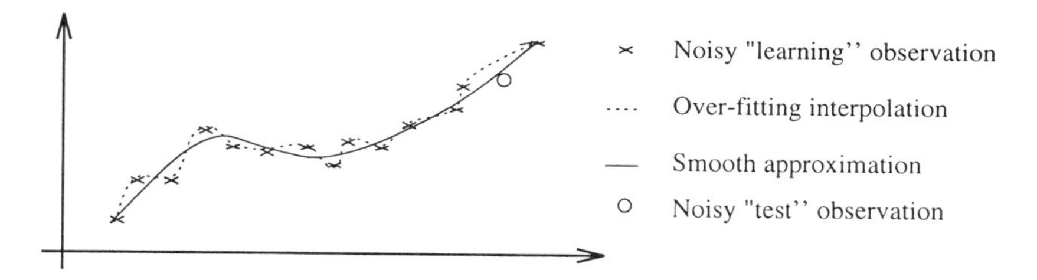

Figure 2.3. Illustration of over-fitting

is proportional to its number of weights, and the complexity of a nearest neighbor rule is proportional to the number of attributes it uses in the distance calculation.

A model which over-fits the training set will be suboptimal in terms of generalization capabilities. Figure 2.3 illustrates the over-fitting phenomenon on a simple one-dimensional curve-fitting problem. For example, in spline approximation[10], if we use a too large number of parameters we may be able to fit the curve to the noisy learning data, but the interpolation/extrapolation on test states may become very poor. One can see that reducing the order of the approximation will actually allow us to improve the approximation.

2.8.1 Bias/variance tradeoff

To explain this over-fitting problem let us briefly discuss in the context of automatic learning the concepts of bias and variance, well-known in statistics.

Using an AL method to derive successive models from successive learning sets (same size, drawn from U) will produce different models, leading to different decision rules. This phenomenon is called *model variance* in the statistical literature : the smaller the learning set size the stronger it is. Furthermore, it is true that if we increase the size of the hypothesis space allowing the AL algorithm to search among more models, variance will increase to some extent.

In addition to variance, most models derived by AL algorithms present also some bias, preventing them from approaching the Bayes rule when the learning set becomes infinitely large. For example, using a linear decision rule to approximate a nonlinear classification boundary leads to bias. Using a decision tree of fixed number of nodes to approximate a complex relationship will also lead to bias. Increasing the space of models allows in general to decrease bias, for example increasing the number of nodes of a decision tree would allow a more accurate approximation.

In other words, if the learning algorithm is well designed, bias will decrease when the model complexity increases. At the same time, the model will depend in a stronger fashion on the random nature of the learning set, thus its variance will increase. Bias

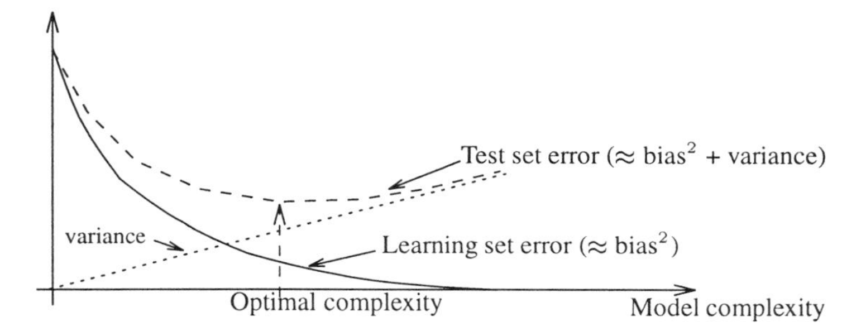

Figure 2.4. Bias/variance tradeoff via cross-validation

and variance lead both to generalization errors and it is necessary to define a tradeoff (see Fig. 2.4).

Thus, the design of "good" learning algorithms consists of two complementary aspects : (i) find model spaces which allow to reduce bias significantly without increasing variance too much; (ii) define the optimal model complexity (leading to an optimal bias/variance tradeoff) for a given learning problem and a given $\boldsymbol{LS}$ size.

In practice, for a given automatic learning method, bias is an increasing function of the physical problem complexity, not of the learning set size. Since variance, on the other hand, is a decreasing function of the learning set size, the optimal model complexity will be an increasing function of both problem complexity and learning set size. While from a qualitative point of view this is well known since the early days of automatic learning, in the last few years theoretical work has lead to quantitative characterizations within different theoretical frameworks [Wol94, Vap95] a detailed discussion of which falls out of the scope of this book.

An important consequence of our discussion is that if we know that a problem is simple (e.g. linear) we should decrease the model space size (e.g. use only linear models) since it will allow us to reduce variance without increasing bias. Thus, we should use the nonlinear techniques such as multilayer perceptrons only if we have good reasons to believe that our problem is indeed nonlinear, and if we have enough data to ensure low variance.

2.8.2 Cross-validation against over-fitting

While theory explains, at least qualitatively, the nature of the over-fitting problem, in practice there are many different ways to fight against it. Some are specific to a particular type of methods, some others are generic. One generic approach, the so-called cross-validation technique, consists of learning a sequence of models of growing complexity, then using an independent test set to evaluate their generalization capabilities and finally selecting the best one. This procedure is illustrated in Fig. 2.4.

Its main drawbacks are computational overhead (one to two orders of magnitude) and data greediness.

Cross-validation may also be used in order to prune models of large complexity. The pruning starts with a very complex model, which is supposed to be over-parameterized. Then, a sequence of models of decreasing complexity is obtained by progressively simplifying this model. E.g. decision trees are pruned by replacing test nodes by terminal ones [BFOS84, Weh93a] ; multilayer perceptrons are pruned by removing some connections (i.e. setting some weights to zero, and/or removing some neurons) and adapting the remaining weights [Hay94] . Cross-validation is used to select among the models of decreasing complexity the best one.

Notice that while cross-validation based model complexity selection works in many cases, it may fail in some cases or, in other cases it may not be the most effective way to control complexity (in terms of both computational burden and resulting model accuracy). For example, in Chapter 5 we will see that the stop-splitting criterion used in decision trees is a very efficient and direct way to control the complexity, which doesn't require a tedious cross-validation.

Let us also mention two other generic approaches to reduce over-fitting. The first one is *regularisation* which is is mainly used in the context of regression. It consists of modifying the square error criterion in order to penalize models which are not smooth enough, e.g. by adding a term proportional to the model curvature [GJP95] . The second one is *model averaging* which consists of building several models and aggregating their outputs in the form of an average, thus reducing variance. In particular, Bayesian averaging uses the Bayesian framework to compute posterior model probabilities and averages models according to these latter [BW91, Bun92, Weh97a] .

2.8.3 Parameter variance

In contrast to model variance discussed above, which relates to the dependence of predictions on the $\boldsymbol{LS}$ used, *parameter variance* relates to the dependence of the model's parameters derived by AL. Examples of such parameters are thresholds used to formulate decision trees and weights in multilayer perceptrons. If a model is used as a black box then the main concern is model variance. However, in many applications, specially in data mining, it is necessary to interpret models produced by automatic learning, for example in order to compare them with prior information and for validation. In this context parameter variance becomes an important concept, and techniques able to estimate it or to reduce it are highly relevant.

In Chapter 5, we will discuss and illustrate parameter variance of decision trees, which are one of the main tools used for data exploration and interpretation. It is also worth mentioning that within the context of parametric statistical methods, discussed in the next chapter, there exist analytical ways to estimate parameter variance.

2.9 SUMMARY AND FURTHER READING

In this chapter we introduced the main terminology and concepts relevant in AL, independently of the AL paradigm and algorithms that will be discussed in the next chapters. By doing so we have stressed upon the similarities among problems and methods rather than on their differences. Although the taste of this chapter may appear to be rather theoretical, we didn't discuss any theoretical approaches to study AL. This is not to say that no such frameworks exist, but their detailed discussion would not be possible with sufficient rigor without involved mathematical derivations, which are not the purpose of this book.

However, much theoretical work was carried out in the recent years and the field of AL learning is less empirical than it looks at first sight. But it is still not quite a science, since several theories are coexisting and unification is not expected in the near future. Moreover, many questions have not been yet addressed at all by any of these theories. Nevertheless, in this section we would like to make some short comments on the different viewpoints adopted. At the end of the chapter we provide a short list of the main references for further reading on this very active research topic.

Statistical learning theory. The recent book by Vapnik [Vap95] gives a very recommendable introduction to this subject. It shows how a general asymptotic learning theory may be obtained by extending the law of large numbers to functional spaces and how non-asymptotic theories may be derived for finite sample sizes using the notions of VC-entropy[11] and VC-dimension, measuring in some sense the complexity of model spaces. The main strength of Vapnik's framework is that it allows to formulate the bias/variance tradeoff in quantitative terms.

Computational learning theory. This approach, proposed by Valiant [Val84] has emerged from the application of computational complexity ideas to AL, yielding the notion of PAC-learnability (probably approximately correct learnability). In addition to deriving error bounds of automatic learning methods, this framework also addresses the computational complexity of the learning algorithms themselves.

Bayesian approach to learning theory. The Bayesian approach to learning consists of defining prior probabilities on the hypothesis space and using learning sets in order to compute their posterior probabilities [Bun92, BW91]. The latter may be exploited either to select a particular model (e.g. the one maximizing the posterior probability) or to achieve model averaging. The general idea behind this is that prior probabilities of individual models will decrease as the hypothesis space increases in size, thus leading to less sharp posterior distributions. It can be shown that this phenomenon is strong enough to naturally lead to some bias/variance tradeoff. In principle the Bayesian framework is stronger than the preceding ones, since it allows one to take into account prior information in the form of prior probabilities of models. However, in practice these latter are seldom available and must be chosen in a more or less pragmatic way. In §3.6.3 we will provide an example of the Bayesian approach in the context of unsupervised learning. In §5.6 we will comment on Bayesian tree averaging.

Minimum description length principle. This idea was introduced first by Rissanen [Ris78]. It exploits the notion of algorithmic complexity of a data string, which is the length of the shortest Turing-machine program which can generate the data string. Algorithmic complexity measures in an absolute way the degree of randomness of a string (up to an additive constant depending on the Turing machine used). For example, if we are unable to write a short program to generate a given data string then we can consider that this string is rather random. Since any data structure may be represented as a string, Turing machine programs may in principle be used both to measure the complexity of a $\boldsymbol{LS}$ and of models constructed by AL.

Thus, the *minimum description length principle* (MDL principle) consists of choosing among all candidate models the one which together with the encoding of its prediction errors on the $\boldsymbol{LS}$ leads to a minimum algorithmic complexity. Increasing the model space will allow us to find better models, eventually making no prediction errors on the $\boldsymbol{LS}$, which would reduce the description length of prediction errors to zero. However, increasing the model space will lead to more candidate models thus increasing the description length of the model itself. The sum of the two terms will lead to a compromise which expresses a particular bias/variance tradeoff.

Towards unification of learning theories. The above frameworks (with some others) are of course interrelated and have their respective limitations. The most recent trend in AL theory focuses on the unification and generalization of these theories. But, for the time being, we are not so far yet and much work remains to be done [Wol94].

Empirical methods. In spite of the advances in theory, much of the research in automatic learning is still empirical. Furthermore, data mining on an actual data base is intrinsically an empirical process, where the analyst applies various algorithms in order to find some interesting results. We recommend very much the recent book by P. Cohen [Coh95], which gives an in depth presentation of empirical methods in AI.

Notes

1. In the literature the terms *model*, *hypothesis* or *pattern* are used to denote the information extracted by automatic learning.

2. Many real-life DBs contain significant amounts of *missing attribute values*, which need to be estimated in some way from the available information. In this book we will only discuss this problem in the context of decision tree induction. Let us note, however, that estimating missing values can be formulated as a supervised learning problem and solved by some of the techniques discussed in this book.

3. The simple *object-attribute-value* representation used throughout this text is the most common representation language used in automatic learning, though not the only one. The more powerful first order predicate calculus is for example used in *inductive logic programming* (ILP) [MD94].

4. In the literature *classification* is used with two different meanings. In the case of *unsupervised learning* one looks at a set of data points and tries to discover classes or groups of similar points; we will use the term *clustering* rather than classification, to denote *unsupervised learning*. In the case of *supervised learning* one is given a set of pre-classified data points and tries to discover a rule allowing one to mimic as closely as possible the observed classification.

5. In the machine learning literature, the term *concept* is also used to denote a class of objects.

6. For the sake of simplicity we assume that in a data base objects are pre-classified with absolute certainty. However, the framework and methods can be easily extended to probabilistic class labels, i.e. cases where each object would be characterized by a vector of class-probabilities. Similarly, instead of using objects described by attribute values, it would be possible to extend the framework to use attribute-value probability distributions. Note also that, while we assume that the data base is described in a deterministic way, models extracted by automatic learning may easily exploit and provide probabilistic information.

7. There is no loss in generality to assume an identical decision and classification space. In particular, some of the classes C_i may be empty, while corresponding to non-empty decisions regions D_i, and vice versa. This allows to represent reject options and to distinguish among sub-categories of classification errors.

8. In order to denote functions we normally use the notation $f(\cdot), g(\cdot) \ldots$ However, when misinterpretation is not possible, we will shorten it to $f, g \ldots$

9. Unless specified otherwise, logarithms are computed in base 2.

10. Spline approximation fits a curve by piecewise quadratic or cubic functions.

11. VC stands for Vapnik-Chervonenkis.

References

[Bun92] W. L. Buntine, *Learning classification trees*, Statistics and Computing **2** (1992), 63–73.

[BW91] W. L. Buntine and A. S. Weigend, *Bayesian back-propagation*, Complex Systems **5** (1991), 603–643.

[Coh95] P. R. Cohen, *Empirical methods in artificial intelligence*, MIT Press, 1995.

[Ris78] J. Rissanen, *Modeling by shortest data description*, Automatica **14** (1978), 465–471.

[Val84] L. G. Valiant, *A theory of the learnable*, Communications of the ACM **27** (1984), no. 11, 1134–1142.

[Vap95] V. N. Vapnik, *The nature of statistical learning theory*, Springer Verlag, 1995.

[Wol94] D. H. Wolpert (ed.), *The mathematics of generalization*, Addison Wesley, 1994, Proc. of the SFI/CNLS Workshop on Formal Approaches to Supervised Learning.

3 STATISTICAL METHODS

3.1 OVERVIEW

Before starting with the description of the statistical methods, we would like to stress the fact that the distinction between statistical and neural network approaches becomes less and less relevant. This will become more obvious in the course of this and the next chapter. One of the common characteristics of these methods is that they handle input information in the form of numerical attributes. Thus all non-numerical information must be translated into numbers via an appropriate coding scheme. In the case of power system problems, this concerns mainly the topological information, which is assumed to be coded by binary 0/1 indicators.

Overall, the statistical approach to AL consists of three conceptual steps : (i) probabilistic description of a class of regression or classification problems by a hypothesis space of joint probability distributions $p(\boldsymbol{a}, \boldsymbol{y})$ or $p(\boldsymbol{a}, c)$, as appropriate, and formulation of simplifying assumptions, if any, about the structure of the underlying probabilistic process; (ii) direct or indirect estimation of the conditional probability distributions $p(\boldsymbol{y}|\boldsymbol{LS}, \boldsymbol{a})$ (resp. $p(c|\boldsymbol{LS}, \boldsymbol{a})$) of the output variables $\boldsymbol{y}$ (resp. c) given the learning set information and the value of the attribute vector of a new observation; (iii) use of the latter model for decision making.

In this framework, the central problem is the estimation of probability distributions from the $\boldsymbol{LS}$, to which there are two broad categories of approaches [DH73] : maximum likelihood (ML) estimators and Bayesian estimators. The ML approach consists of choosing the hypothesis which maximizes the likelihood of observing the $\boldsymbol{LS}$. On the other hand, the Bayesian approach proceeds as follows :

1. assumption of prior probabilities of all candidate hypotheses;

2. computation of posterior probabilities of hypotheses given the $\boldsymbol{LS}$;

3. exploitation of the posterior probabilities, either to select the most probable hypothesis given the learning set, or to compute an average hypothesis obtained by weighting the predictions of individual ones by their posterior probability.

Much of the earlier work in statistics has concentrated on determining analytical conditions such as independence and linearity under which such estimators can be derived in a direct analytical way; this is the realm of so-called *parametric statistics*. However, since the advent of digital computers, many new methods have been proposed which can be used when the simplifying conditions do not hold, yielding so-called *non-parametric or distributionfree statistics*. These latter methods are closely related, sometimes strikingly similar, to methods proposed in the artificial neural network and machine learning communities. Both kinds of methods may be useful in practice and it is often a good idea to start with the simpler parametric ones and call for the more sophisticated non-parametric techniques only if necessary.

Thus in this chapter we pursue a threefold objective : (i) to give an overview of what kind of AL methods statistics can provide; (ii) to describe a selection of methods complementary to those which will be discussed in the next two chapters; (iii) to discuss practically most relevant methods and highlight their strengths and limitations regarding large scale problems such as those found in many power system applications. In our presentation we try to minimize mathematical derivations and proofs which can be found in the references given at the end of this chapter.

A final comment concerns data preprocessing. All AL methods may take advantage or even need some degree of data preprocessing, like attribute selection, normalization, de-correlation .... Since some the AL methods not yet presented may help to preprocess a data base before applying another one, we will postpone our discussion of these techniques to Chapter 6.

3.2 PARAMETRIC METHODS

Below we discuss a subset of parametric methods, mainly linear models. The reader should be aware of the fact that there are different approaches to yield linear classification models, which provide essentially the same result in the ideal conditions but behave differently when these conditions are not satisfied.

3.2.1 Classification

Continuous input space. Let us first suppose that attributes are real numbers and that their class conditional probability distributions are Gaussian. Namely

$$p(\boldsymbol{a}|c_i) \triangleq \frac{1}{(2\pi)^{\frac{n}{2}}|\Sigma_i|^{\frac{1}{2}}} \exp\left[-\frac{1}{2}\left(\boldsymbol{a}(o)-\boldsymbol{\mu}_i\right)^T \Sigma_i^{-1}\left(\boldsymbol{a}(o)-\boldsymbol{\mu}_i\right)\right], \tag{3.1}$$

where $\boldsymbol{\mu}_i \triangleq E\{\boldsymbol{a}|c_i\}$ denotes the mean attribute vector in class i and Σ_i is the class-conditional covariance matrix.

Then, the Bayes decision rule - yielding minimum error rate - is obtained by choosing the class c_j such that the value of its posterior probability

$$P(c_j|\boldsymbol{a}) = \frac{p(\boldsymbol{a}|c_j)p_j}{p(\boldsymbol{a})} \tag{3.2}$$

is maximal, or equivalently, such that

$$\log P(c_j|\boldsymbol{a}) = \log p(\boldsymbol{a}|c_j) + \log p_j + \log p(\boldsymbol{a}) \tag{3.3}$$

is maximal.

Since $p(\boldsymbol{a})$ is independent of the class c_j, this is equivalent to maximizing

$$g_j(\boldsymbol{a}(o)) \triangleq \log p(\boldsymbol{a}(o)|c_j) + \log p_j \tag{3.4}$$

$$= -\frac{1}{2}\left(\boldsymbol{a}(o)-\boldsymbol{\mu}_j\right)^T \Sigma_j^{-1}\left(\boldsymbol{a}(o)-\boldsymbol{\mu}_j\right) - \frac{1}{2}\log|\Sigma_j| + \log p_j + \frac{n}{2}\log 2\pi, \tag{3.5}$$

where we can drop the term $\frac{n}{2}\log 2\pi$ which is independent of the class.

Thus, to each class c_i corresponds a quadratic function $g_i(\boldsymbol{a}(o))$, and the equalities $g_i = g_j$ hence lead to quadratic discriminants in the general case.

However, these quadratic hyper-surfaces degenerate into linear hyper-planes when the class-conditional covariance matrices are identical. Indeed, let us consider the two-class case, where $\Sigma_1 = \Sigma_2 = \Sigma$. Then the Bayes optimal decision rule is to decide class c_1, whenever $g_1(\boldsymbol{a}) - g_2(\boldsymbol{a}) > 0$, namely when

$$\boldsymbol{a}^T(o)\Sigma^{-1}(\boldsymbol{\mu}_1-\boldsymbol{\mu}_2) > \tfrac{1}{2}(\boldsymbol{\mu}_1+\boldsymbol{\mu}_2)^T\Sigma^{-1}(\boldsymbol{\mu}_1-\boldsymbol{\mu}_2) + \log\tfrac{p_2}{p_1}. \tag{3.6}$$

In practice, the quadratic discriminant (or its linear degenerate) may in principle be directly determined by estimating the class-conditional mean vectors and covariance matrices and substituting in the above formula. However, it should be noticed that estimates of covariance matrices are generally ill-conditioned as soon as the data departs from the ideal assumptions. Furthermore, even in the ideal situation of Gaussian distributions, as soon as the number of attributes becomes important (say more than 30) the number of learning states required to provide an acceptable accuracy becomes prohibitive.

Fisher's linear discriminant. Given these limitations, Fisher [Fis36] proposed a more robust also direct technique, which in the ideal "identical-covariance" case provides the same model but can also be applied when the covariance matrices are different.

The basic idea behind the Fisher's linear discriminant is to replace the multi-dimensional attribute vectors by a single feature resulting from a linear transformation of the attributes. Thus the objective is to define a linear transformation maximizing the separation of objects of different classes, on the new feature axis.

Let us define the class conditional means of the attribute vector by

$$\overline{\boldsymbol{a}}_i \triangleq \sum_{o \in LS; c(o)=c_i} \boldsymbol{a}(o), \tag{3.7}$$

and the class conditional scatter of the linear projection on vector $\boldsymbol{w} = (w_1, \ldots, w_n)^T$ of the samples, by

$$\tilde{s}_i^2 \triangleq \sum_{o \in LS; c(o)=c_i} (\boldsymbol{w}^T \boldsymbol{a}(o) - \boldsymbol{w}^T \overline{\boldsymbol{a}}_i)^2. \tag{3.8}$$

Then *Fisher's linear discriminant* between two classes is defined by the weight vector $\boldsymbol{w}^*$ maximizing the following criterion function

$$J(\boldsymbol{w}) \triangleq \frac{(\boldsymbol{w}^T \overline{\boldsymbol{a}}_1 - \boldsymbol{w}^T \overline{\boldsymbol{a}}_2)^2}{p_1 \tilde{s}_1^2 + p_2 \tilde{s}_2^2}, \tag{3.9}$$

which is the ratio of the distance from the projected mean vectors to the mean class-conditional standard deviation of projected feature values. (In the pure Fisher's linear discriminant, the classes are supposed to be equiprobable.)

Let us also define the mean class-conditional sample covariance matrix $\hat{\Sigma}_W$ by

$$\hat{\Sigma}_W \triangleq p_1 \hat{\Sigma}_1 + p_2 \hat{\Sigma}_2, \tag{3.10}$$

where the matrices $\hat{\Sigma}_i$ are the sample estimates of the class conditional covariance matrices, obtained by

$$\hat{\Sigma}_i \triangleq \frac{1}{n_{i.}} \sum_{o \in LS; c(o)=c_i} [\boldsymbol{a}(o) - \overline{\boldsymbol{a}}_i] * [\boldsymbol{a}(o) - \overline{\boldsymbol{a}}_i]^T, \tag{3.11}$$

where $n_{i.}$ denotes the number of learning states of class i.

Then, it may easily be shown that an explicit form is obtained for $\boldsymbol{w}^*$ by the following formula

$$\boldsymbol{w}^* = \hat{\Sigma}_W^{-1} (\overline{\boldsymbol{a}}_1 - \overline{\boldsymbol{a}}_2), \tag{3.12}$$

provided that the matrix $\hat{\Sigma}_W$ is non-singular. Otherwise, the optimal direction may be determined by an iterative gradient descent least squares technique [DH73].

To obtain a decision rule for class 1, in the form $g(\boldsymbol{a}(o)) = \boldsymbol{w}^T\boldsymbol{a}(o) + w_0 \geq 0$, in addition to choosing $\boldsymbol{w}$ it is required to define an appropriate threshold w_0. In the standard Fisher's linear discriminant method, this threshold is chosen directly on the basis of the distribution parameters, in the following way :

$$w_0 \triangleq -\frac{1}{2}(\overline{\boldsymbol{a}}_1 + \overline{\boldsymbol{a}}_2)^T \hat{\Sigma}_W^{-1}(\overline{\boldsymbol{a}}_1 - \overline{\boldsymbol{a}}_2) + \log\frac{p1}{p2}. \tag{3.13}$$

However, once the vector $\boldsymbol{w}^*$ has been fixed, it is a simple scalar optimization problem to choose the appropriate value of w_0. Therefore, it may be done easily, e.g. using a simple enumerative search, similar to the one discussed in Chapter 5 to search for optimal thresholds in decision tree induction.

Binary input space. Another situation where the linear model is optimal corresponds to the case where attributes are binary and class-conditionally independent.

Indeed, when the attributes are independent given the class, their conditional probability can be factorized as

$$P(\boldsymbol{a}|c_j) = \prod_{i=1,n} P(a_i|c_j). \tag{3.14}$$

Further, if the attribute values are binary, they may be encoded as 0/1 values and $P(c_j|\boldsymbol{a})$ may be rewritten as

$$P(c_j|\boldsymbol{a}) = \frac{p_j \prod_{i=1,n} P(a_i = 1|c_j)^{a_i} P(a_i = 0|c_j)^{1-a_i}}{P(\boldsymbol{a})}. \tag{3.15}$$

Again, we may drop the class independent denominator and take logarithms to yield linear discriminants.

Notice that the generalization of the above idea to non-binary class-conditionally independent features yields the so-called "naive Bayes" classifier, often misnamed as the Bayes rule (see §3.5.3).

3.2.2 Regression

Let us consider a hypothesis space $\mathcal{R}$ of homogeneous linear regression models in the form

$$r = \boldsymbol{w}^T\boldsymbol{a}. \tag{3.16}$$

Then, given a $\boldsymbol{LS}$ of pairs $(\boldsymbol{a}^i, y^i)$ the learning problem in the least-squares setting amounts to searching vector $\boldsymbol{w}^*$ which minimizes the square error of eqn. (2.41). It is well known that this problem has a direct solution, based on the normal equations. Denoting by $A = (\boldsymbol{a}^1, \ldots, \boldsymbol{a}^N)$ the $\boldsymbol{LS}$ input matrix, and by $\boldsymbol{y} = (y^1, \ldots, y^N)^T$ the vector of corresponding output values the solution is given by

$$\boldsymbol{w}^* = (AA^T)^{-1}A\boldsymbol{y}. \tag{3.17}$$

Notice that eqn. (3.16) supposes that both attribute values and outputs have zero mean (thus, their mean must be subtracted before applying (3.16)). Under these conditions, the matrix (AA^T) is actually the sample covariance matrix multiplied by N. If, further, the attribute values have been normalized by dividing them by their standard deviation, (AA^T) is equal to the sample correlation matrix multiplied by N.

In the case where the output variable y is indeed a linear function of the inputs up to an independent zero mean noise term, it is true that the solution of eqn. (3.17) provides an unbiased estimator of the true output y. Further, among all unbiased estimators it has minimal variance and if the noise term has Gaussian structure, it maximizes the likelihood of the sample.

Note that (AA^T) should be non-singular (and well conditioned) in order to allow solving eqn. (3.17). This will not be the case if there are linear (or almost linear) relationships among attributes. Furthermore, if the learning sample size is small compared to the number of attributes, the resulting model may have high variance. In either of these circumstances, it is possible to reduce variance at the expense of some bias, for example by dropping the least useful variables, or better, by using a small number of principal components. Thus, even in such a simple case where the problem is linear, the optimum model corresponding to the best bias/variance tradeoff would not be the one given by eqn. (3.17).

3.2.3 Discussion

The preceding discussion aimed at showing that under some well defined conditions the solution to the AL problem leads to simple parametric models whose parameters may be estimated using direct techniques from linear algebra. One of the main difficulties with these techniques is that they often require the estimation of inverse covariance matrices. These are ill-conditioned in many cases when there are close to linear relationships among variables. And, as soon as the number of attributes becomes large, these estimates have a high variance even for large samples. Thus, indirect iterative optimization techniques have been proposed in the literature which aim at solving the problem of minimizing the apparent error without calling for the estimation of covariance matrices and their inverses. Indeed, if we are searching for a linear model containing n or $n+1$ parameters why should we bother in estimating matrices containing $n(n+1)$ terms.

It turns out that using iterative optimization techniques, such as those discussed in the next section and in Chapter 4, it is indeed possible to make better use of the information contained in the learning set and avoid some of the pitfalls discussed above. Nevertheless, the simple direct methods above are useful tools because they are fast and lead to direct, unique solutions.

Generalized linear models are obtained if we extend the attribute space by including some attributes defined as nonlinear functions of the original ones and applying the above kind of technique in the extended space. By doing so, one can induce a larger hypothesis space and possibly find better solutions to a larger class of AL learning problems. However, as always, there is a price to pay : increasing the hypothesis space leads to a larger number of parameters and increases model variance. Thus, whether the resulting nonlinear solution will be better than the linear one will strongly depend both on the problem complexity and on the $\boldsymbol{LS}$ size.

Illustration. Figure 3.1 illustrates these ideas on the OMIB transient stability problem introduced in Chapter 1. The left hand scatter plot shows the relation between the CCT and its linear estimate in terms of Pu and Qu; the right hand diagram shows the improvement obtained with a quadratic approximation in the form

$$r = w_0 + w_1 Pu + w_2 Qu + w_3 Pu^2 + w_4 Qu^2 + w_5 PuQu.$$

Note that using the first model to classify states with respect to the threshold of 155ms yields a test set error rate of 2.5% ; using the quadratic model yields an error rate of 0.65%. Note also that the quadratic approximation is less accurate than the MLP of Chapter 1, but better than the nearest neighbor regression. Moreover, the learning algorithm is about 300 times faster than the MLP learning.

Such generalized linear models together with iterative indirect optimization techniques actually bridge the gap between parametric and non-parametric methods. Although they certainly have much potential, they have become less popular in the recent years, in particular due to the emergence of the neural network approaches described in Chapter 4 and the projection pursuit techniques discussed in the next section.

3.3 PROJECTION PURSUIT

The projection pursuit regression technique models a regression function $\boldsymbol{r}(\cdot)$ as a linear combination of smooth functions of linear combinations (or projections) of the attribute values. Thus, the model assumes the following formulation

$$\boldsymbol{r}(\boldsymbol{a}) \triangleq \overline{\boldsymbol{y}} + \sum_{i=1,K} \boldsymbol{v}_i f_i(\boldsymbol{w}_i^T \boldsymbol{a}), \tag{3.18}$$

where the order K, the r-vectors $\boldsymbol{v}_i$, the n-vectors $\boldsymbol{w}_i$ and the scalar functions $f_i(\cdot)$ are determined on the basis of the learning set, in an iterative attempt to minimize the

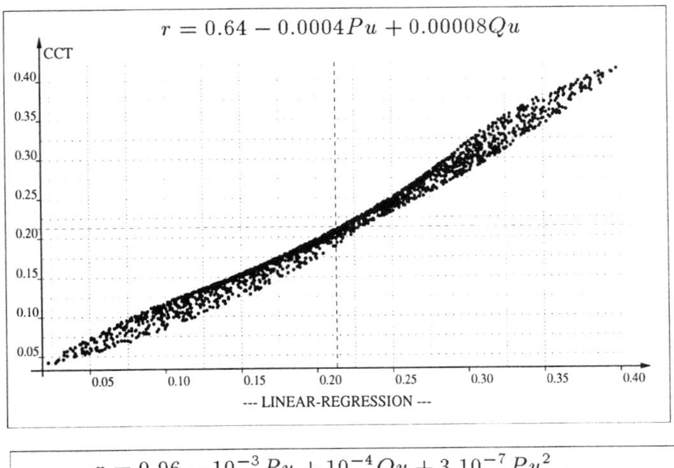

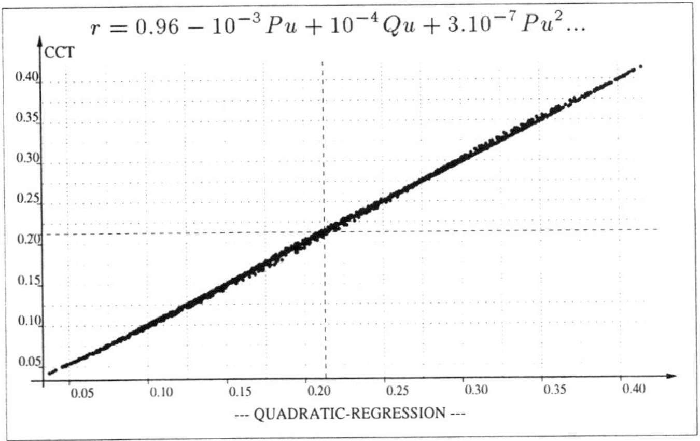

Figure 3.1. Illustration of linear vs quadratic regression on the OMIB data base

mean square error

$$SE \triangleq N^{-1} \sum_{o \in LS} ||\boldsymbol{y}(o) - \boldsymbol{r}(\boldsymbol{a}(o))||^2. \tag{3.19}$$

For classification problems, the standard class-indicator encoding is used, which is defined by

$$y_i(o) = \delta_{c(o),c_i}, \; \forall \; i = 1, \ldots, m. \tag{3.20}$$

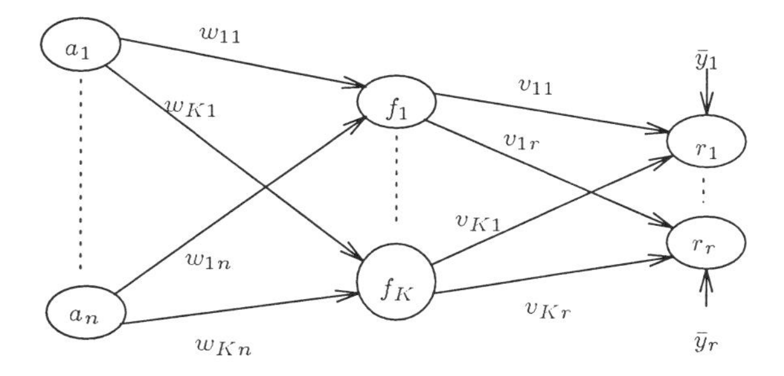

Figure 3.2. Graphical representation of the projection pursuit model

In the basic approach the functions f_i are special scatter plot smoothers, normalized in the following way

$$\sum_{o \in LS} f_i(\boldsymbol{w}_i^T \boldsymbol{a}(o)) = 0 \text{ and } \sum_{o \in LS} f_i^2(\boldsymbol{w}_i^T \boldsymbol{a}(o)) = 1, \tag{3.21}$$

and the projection vectors $\boldsymbol{w}_i$ are normed :

$$\sum_{j=1,n} w_{ij}^2 = 1. \tag{3.22}$$

The striking similarity of this model with a single hidden layer MLP is shown in Fig. 3.2. However, the originality of the projection pursuit regression technique is that both model complexity (the order K) and the smooth activation functions $f_i(\cdot)$ are determined on the basis of the learning set data, while in the basic multilayer perceptron they are chosen a priori by the user, which can lead to overly complex structures with many redundant parameters.

Forward growing of projection pursuit. At each step j of the procedure, the order of the model is increased by one unity, by adding an additional projection direction $\boldsymbol{w}_j$ and smooth function f_j and determining the vector $\boldsymbol{v}_j$. During this first step, the parameters of the preceding directions are kept constant.

The second step consists of adjusting in a back-fitting approach all the parameters of all directions $k \leq j$ in a cyclic fashion, so as to minimize the square error (3.19).

Finally, the model growing procedure stops when the square error is sufficiently low or when it does not improve anymore.

A complementary approach to the above growing consists of generating the models in decreasing order of their complexity K, by starting with a sufficiently high value of K and pruning at each step the least active part of the model, corresponding to the projection direction which influences the least strongly the output values. This is defined as the direction i which minimizes the sum

$$I_i \triangleq \sum_{j=1,r} |v_{ij}|. \tag{3.23}$$

Back-fitting. The heart of the algorithm consists of back-fitting a group of parameters, $\boldsymbol{w}_i$, f_i, and $\boldsymbol{v}_i$, corresponding to one of the current projection directions $i \leq j$. This is done in an iterative fashion.

1. Adjusting $\boldsymbol{v}_i$ can be done in a direct fashion by setting the derivatives of the SE to zero with respect to each component of $\boldsymbol{v}_i$. This yields a linear equation, since the SE is quadratic in $\boldsymbol{v}_i$.

2. To adjust the smooth functions $f_i(\cdot)$, we proceed in two steps. First, non-smooth function values $f_i(\boldsymbol{w}_i^T \boldsymbol{a}(o))$ are determined for each object $o \in \boldsymbol{LS}$. Again, since the SE is quadratic in f_i, this can be done in a direct linear computation, setting the partial derivatives of the SE w.r.t. $f_i(\boldsymbol{w}_i^T \boldsymbol{a}(o))$ ($\forall\ o \in \boldsymbol{LS}$) to zero. Second, the resulting "optimal" values

$$\left(\boldsymbol{w}_i^T \boldsymbol{a}(o), f_i^*(\boldsymbol{w}_i^T \boldsymbol{a}(o))\right), \ \forall\ o \in LS, \tag{3.24}$$

are used as target values to determine the smooth interpolation function. For a further discussion of various alternative schemes for this unidimensional smoothing, we refer the interested reader to [HYLJ93, SW96].

3. Finally, to adjust the projection direction $\boldsymbol{w}_i$, an iterative gradient descent or Newton method should be used, since the SE is not a quadratic function of $\boldsymbol{w}_i$.

Discussion. One of the advantages of the *projection pursuit* regression method with respect to standard feed-forward neural network techniques lies in the greater simplicity of the resulting structure. This is due to the automatic determination of the neuron activation function together with the adaptation of the model complexity to the data. While similar neural network growing techniques have been proposed in the literature, the projection pursuit approach has been found to be superior in performance to the cascade correlation techniques proposed by Fahlman and Lebière for neural networks [FL90]. Actually, the main motivation of cascade correlation is to increase the speed of learning convergence and not so much to improve the model accuracy.

Admittedly, in high dimensional attribute spaces the projection directions found by this method may become difficult to interpret. Thus, Friedman and Stuetzle have proposed various extensions to the basic method to improve its data exploration features [FS81]. For example, by restricting the number of attributes combined in any projection, the method may provide interesting two or three dimensional directions for data exploration. With these extensions this method would provide similar features to decision tree induction, with the additional capability of providing a *smooth* nonlinear input/output modeling capability, which would be particularly interesting for the estimation of power system security *margins*.

Finally, we mention that the SMART implementation of the projection pursuit regression technique was applied, in the context of the StatLog project, on two power system security related data sets [MST94]. In both cases this method scored best in terms of reliability (but also slowest in terms of learning CPU time). This, in addition to the possibility of exploiting the continuous security margins, provides a strong motivation for further exploration of the capabilities of these projection pursuit approaches in the context of power system security problems. We refer the reader to Table 10.1 were the results of this method are compared with other AL methods, on a data base corresponding to transient stability assessment of the EDF system.

3.4 DISTANCE BASED METHODS

3.4.1 Basic KNN

Nearest neighbor (1NN) methods have been applied both for density estimation, classification and regression. We discuss only the latter two applications.

Classification. Given a learning set $\boldsymbol{LS}$ and a distance Δ_a defined in the attribute space, the nearest neighbor classifier consists of classifying an object o in the class $c(o')$ of the learning state o' of minimal distance, i.e. $o' = \arg\min_{LS} \Delta_a(o, o')$.

Asymptotically, when the $\boldsymbol{LS}$ size $N \to \infty$, the nearest neighbor o' converges towards the object o. Thus, its class $c(o')$ has an expected asymptotic probability of being the correct class $c(o)$ equal to

$$\sum_{i=1,m} p_i(\boldsymbol{a}(o))p_i(\boldsymbol{a}(o)). \tag{3.25}$$

From this, it may be derived that in an m class problem the asymptotic error rate of the nearest neighbor rule is bounded by [DK82]

$$P_e^{Bayes} \leq P_e^{NN} \leq P_e^{Bayes}(2 - \frac{m}{m-1}P_e^{Bayes}), \tag{3.26}$$

where P_e^{Bayes} denotes the error rate of the Bayes rule. Note that these bounds are the tightest bounds which can be derived. Thus, depending on the class-conditional distributions, either bound may be achieved. In particular, in the worst case the nearest neighbor rule may be rather suboptimal.

This sub-optimality of the nearest neighbor method related to the fact that it extrapolates the classification of the $\boldsymbol{LS}$ without any smoothing, thus yielding a high

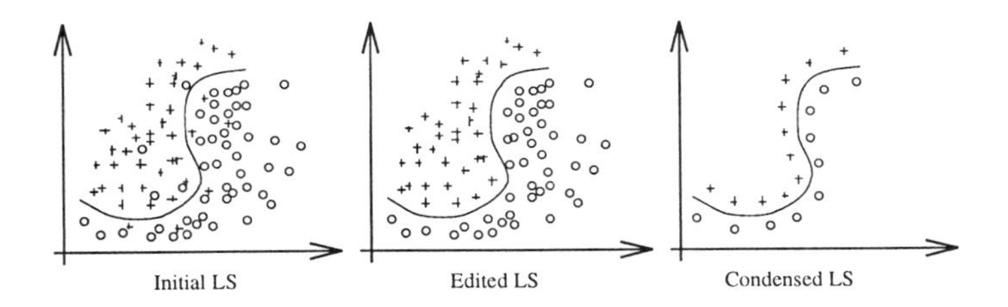

Figure 3.3. Nearest neighbor, editing and condensing

variance. It is interesting to observe that this over-fitting sub-optimality remains a problem even for very large samples.

The first approach to solving this problem consists of reducing the locality of the 1NN information by using more than one nearest neighbor. This leads to the so-called KNN or (K, L)NN rules [DK82].

The basic KNN rule consists of searching for the K nearest neighbors of an attribute vector and estimating the class probabilities by

$$\hat{p}_i(\boldsymbol{a}(o)) \triangleq \frac{n(K, o, c_i)}{K}, \tag{3.27}$$

where $n(K, o, c_i)$ is the number of learning states of class c_i among o's K nearest neighbors.

Asymptotically, the KNN is Bayes optimal, strictly speaking, if the number K increases with N, such that

$$\lim_{N \to \infty} K(N) = \infty \text{ and } \lim_{N \to \infty} \frac{K(N)}{N} = 0. \tag{3.28}$$

Indeed, $\lim_{N \to \infty} \frac{K(N)}{N} = 0$ guarantees that the K nearest neighbors still converge towards the object o in the attribute space (zero bias), while $\lim_{N \to \infty} K(N) = \infty$ guarantees that the class-probability estimates converge towards the true values (zero variance). For a finite learning set size the effect of increasing K, will be to reduce variance of the probability estimates but to increase bias. Thus, there is an optimal bias/variance tradeoff, corresponding to an optimal value of K , depending both on the problem and $\boldsymbol{LS}$ size N.

The second approach to improve the 1NN rule consists of editing the learning set by removing those learning states which are surrounded by states of a different class. This consists of increasing the probability of the nearest neighbor to belong to the majority class, and thus leads to nearly optimal decision rules.

In addition to these editing techniques, condensing algorithms may be used to dramatically reduce the size of the required data base, by removing the states which do not contribute to defining the decision boundary. Figure 3.3 illustrates graphically the editing and condensing techniques, which are discussed in full detail in [DK82]. It should be noted that while these techniques improve error rates and particularly dramatically reduce the CPU times, they unfortunately strongly reduce the locality of the nearest neighbor classifier, which is however a desirable practical feature of the 1NN method, as is discussed below in §3.4.3.

Regression. Another possible use of the nearest neighbor approach is for regression problems. In this case the following type of regression function may, for example, be used

$$r(o) \triangleq \frac{\sum_{o' \in KNN(LS,o)} y(o')\Delta_a^{-1}(o,o')}{\sum_{o' \in KNN(LS,o)} \Delta_a^{-1}(o,o')} \tag{3.29}$$

where we have denoted by $KNN(LS, o)$ the set of the K nearest neighbors of o.

3.4.2 Leave-one-out with KNN

It is interesting to notice that the leave-one-out error estimate (see §2.7.9) of the KNN method may be obtained in a very straightforward way.

Indeed, it amounts to screening through the N learning states, one by one, and searching their K nearest neighbors within the remaining $N - 1$ states.

In terms of computational efficiency, it will require in the order of N^2 distance computations, while test set error rate would require in the order of $M \times N$ distance computations. Thus, as far as M and N are of the same order of magnitude the leave-one-out method will remain competitive. In practice, it can be used in order to determine an appropriate value of K.

3.4.3 Discussion

The nearest neighbor rule is a very simple and easy to implement approach. But is has a main disadvantage, that of requiring a very large number of learning states in order to become robust with respect to the definition of distances. In particular, in the case of high dimensional attribute spaces the method may rapidly require prohibitively large samples. Thus, to be effective it must in general rely on prior feature selection and/or extraction techniques, so as to reduce the attribute space dimensionality.

At the same time, while the learning of the basic nearest neighbor rule merely consists of storing the data base, the complexity of using this information for new classifications is directly proportional to the product of the number N of learning states and the dimension n of the attribute space. This may be several orders of magnitude higher than the time required by competing techniques and only rather sophisticated search algorithms can allow us to reduce the CPU time. Nevertheless, in the context of power system security assessment this would not be a very strong limitation, thanks

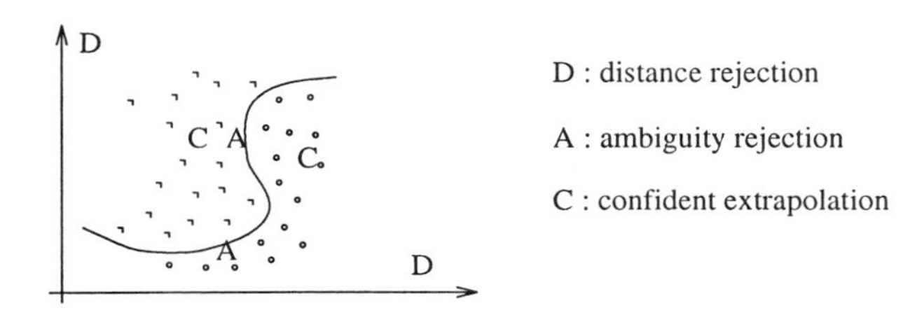

Figure 3.4. Nearest neighbor ambiguity and distance rejection

Table 3.1. Error rates (%) of KNN classifiers

K	1	3	5	7	9
28 attributes	5.92	5.20	4.80	4.06	4.48
3 attributes	4.24	3.12	3.04	2.80	2.96

to the increased CPU speeds and the relatively limited size of data bases to several thousand states. However, in applications like image or printed character recognition, where data bases of several millions of objects are frequent, this becomes one of the main concerns of this method.

The main practical interest of this approach is its ability to identify the objects of a data base which are most similar to an unseen situation and which are used to extrapolate output information. For example, in a security assessment environment, the "nearest" known security scenarios may be supplied to an operator as an explanation or justification of the security assessment of the current system state. In particular, the main differences with the current situation may be analyzed so as to decide whether and how their information may be extrapolated. This could, for example, allow us to use local linear approximation techniques, so as to infer security margins, and to provide rejection options for states either too close to the classification boundary (ambiguity rejection) or too far away from any reference case (distance rejection) (see Fig. 3.4). These possibilities in the context of security assessment applications will be further discussed in Part II of the book.

Illustration. As an illustration of the typical behavior of the KNN method, let us look at a voltage security assessment example discussed in [Weh95a]. Table 3.1 shows the influence of K on the test set error rates obtained for two different sets of attributes. In each case the standard Euclidean distance was used, and the attributes were normalized by dividing their value by their standard deviation (see §6.2.1).

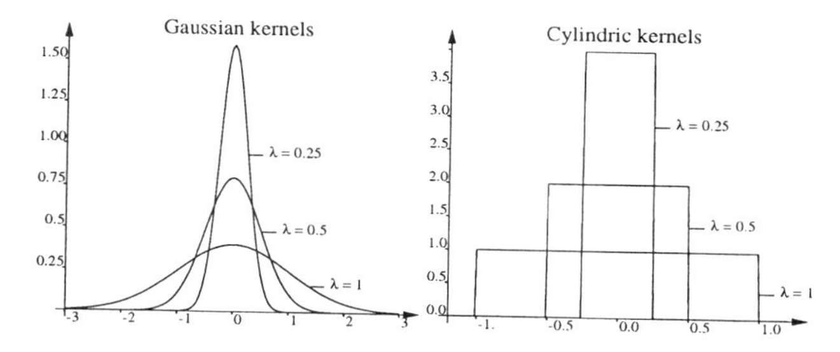

Figure 3.5. Various kernel functions and smoothing parameters

These results illustrate how increasing the value of K allows us to reduce the error rate. They suggest that for both sets of attributes the optimal value of K is equal to 7. It is also interesting to note that reducing the number of attributes has allowed us to significantly improve the performance, both in terms of reliability and efficiency. In this particular example, we have built a decision tree so as to identify among the 28 attributes the 3 most significant ones.

Thus, while the nearest neighbor technique is a very attractive approach, its main weakness stems from its sensitivity to the chosen distance measure. To make it practical it needs to be enhanced by an algorithm able to choose automatically which attributes should be used and how they should be combined in the KNN method for a specific problem. Such a technique, based on the combined use of decision trees and genetic algorithms will be described in Chapter 6.

3.5 OTHER NON-PARAMETRIC TECHNIQUES

Below we group together three other classical non-parametric techniques. Although we do not believe they would be very useful in our practical context (i.e. more accurate or complementary in terms of functionality), we provide a brief description since some of the results discussed later pertain to one of these methods.

3.5.1 Kernel density estimation

While the KNN and projection pursuit methods aim at directly modeling the conditional class-probabilities $p_i(\boldsymbol{a})$, the kernel density estimation approach operates indirectly by providing a non-parametric estimate of the class conditional attribute densities $p(\boldsymbol{a}|c_i)$.

This approach uses the following expansion

$$\hat{p}(\boldsymbol{a}|c_i) \triangleq \frac{1}{n_{i.}} \sum_{o \in LS; c(o)=c_i} \phi(\boldsymbol{a}, \boldsymbol{a}(o), \lambda), \tag{3.30}$$

where the function $\phi(\cdot, \boldsymbol{a}(o), \lambda)$ is a kernel function centered at $\boldsymbol{a}(o)$, and λ is a smoothing parameter. Various kernel functions are suggested in Fig. 3.5 together with the effect of the smoothing parameter.

In addition to different possible choices for the kernel function, discussed in most pattern recognition textbooks [DH73, Han81, DK82], it is also important to choose the smoothing parameter to adapt the method to the data. Actually, it turns out that the choice of the smoothing parameter, which is the kernel density version of our, by now, familiar over-fitting problem, is much more important in practice than the choice of kernel function type.

Various techniques have been proposed to estimate the value of λ on the basis of the data. One possibility consists of maximizing the "leave-one-out" sample likelihood, defined by

$$L(LS|\lambda) \triangleq \prod_{o \in LS} \hat{p}'(\boldsymbol{a}(o)|c(o)) \tag{3.31}$$

where $\hat{p}'(\boldsymbol{a}(o)|c(o))$ is the density estimate at point $\boldsymbol{a}(o)$ for class $c(o)$, obtained when o is removed from the learning set, i.e.

$$\hat{p}'(\boldsymbol{a}(o)|c(o)) \triangleq \frac{1}{n_{i.} - 1} \sum_{o' \in LS; c(o')=c(o); o' \neq o} \phi(\boldsymbol{a}(o), \boldsymbol{a}(o'), \lambda). \tag{3.32}$$

The expression (3.31) may then be optimized with respect to λ by a one dimensional numerical search technique in the semi space $\lambda \in]0 \ldots \infty[$.

3.5.2 Histograms

A very simple approach to non-parametric density estimation is the histogram approach. Basically, this method consists of dividing a priori the attribute space into subregions and counting the relative number of states of each class falling into each subregion. In the simplest case the regions are defined by dividing the range of each interval into a fixed number of regular sub-intervals. The advantage of this approach with respect to kernel density estimation or nearest neighbor is that it does not require to store any of the learning states.

However, in order to make this approach applicable in the case of multidimensional attribute spaces, the size of the elementary regions must be adapted to the learning set representativity, in particular to avoid empty regions and to minimize the variations among neighboring cells. This is the histogram version of the over-fitting problem, for which, not surprisingly, smoothing solutions have been proposed in the literature [Han81]. In spite of these improvements, we believe that the approach is mainly useful in one, two or three dimensional situations.

Illustration. Figure 3.6 illustrates the class-conditional histograms of the two attributes Pu and Qu of the OMIB example, determined on the $\boldsymbol{LS}$. The diagrams

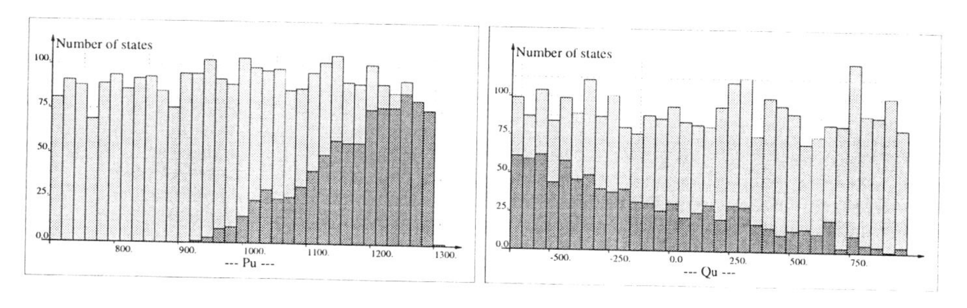

Figure 3.6. One-dimensional histograms for the OMIB data base

illustrate in particular that due to the random sampling the probability density estimates thus obtained are not smooth and suffer from large variance. Notice that this high variance does not necessarily translate into high classification errors [Fri97].

We will see that such one-dimensional histograms may be exploited in the context of the "naive" Bayes approach discussed in the next section.

3.5.3 Naive Bayes

We already mentioned that an interesting situation occurs when the attributes are independently distributed in each class. Then the class conditional probability densities may be factorized in the following way

$$p(\boldsymbol{a}|c_i) = \prod_{j=1,n} p(a_j|c_i), \tag{3.33}$$

and the multi-dimensional estimation problem reduces to $m \times n$ one-dimensional density estimations, $p(a_j|c_i)$ for each attribute.

In particular, for discrete attributes this amounts to counting the number of occurrences of each possible attribute value in each class and using these values in order to estimate the corresponding probabilities in a maximum likelihood or Bayesian approach.

For real valued attributes either a parametric estimator or one of the above non-parametric techniques (KNN, kernel density estimation or histograms) may be used in order to estimate the scalar probability density function $p(a_i|c_i)$. The most straightforward non-parametric technique is in this particularly simple one-dimensional case the histogram approach, due to its simplicity and computational efficiency.

It is interesting to notice that the above naive Bayes classifier leads to generalized linear discriminant functions. Indeed, replacing the probabilities $p(a_j|c_i)$ by their

Table 3.2. Attribute statistics of the OMIB problem

Class	p_i	Pu		Qu	
		μ	σ	μ	σ
Secure	0.697	926	137	280	460
Insecure	0.303	1178	85	-127	401

estimate $\hat{p}(a_j|c_i)$ and taking the logarithm yields

$$\log p(c_i|\boldsymbol{a}) = \log p_i + \log p(\boldsymbol{a}|c_i) = \log p_i + \sum_{j=1,n} \log \hat{p}(a_j|c_i) + \text{Cnst}. \tag{3.34}$$

Illustration. Coming back to our OMIB example, let us see how the naive Bayes approach works. Table 3.2 indicates the mean and standard deviations of the two attributes Pu and Qu both in the secure and in the insecure classes, as estimated from our $\boldsymbol{LS}$, as well as the estimated prior class probabilities. Notice that mean values indicate that unstable states correspond to high values of Pu and low values of Qu.

Assuming that Pu and Qu are independently distributed in each class according to a Gaussian distribution, will yield the following for the secure class

$$\log P(Sec|Pu, Qu) = \log \frac{0.697}{2\pi 137 * 460} - \frac{(Pu - 926)^2}{2 * 137} - \frac{(Qu - 280)^2}{2 * 460}, \tag{3.35}$$

and

$$\log P(Ins|Pu, Qu) = \log \frac{0.303}{2\pi 85 * 401} - \frac{(Pu - 1178)^2}{2 * 85} - \frac{(Qu + 127)^2}{2 * 401} \tag{3.36}$$

for the insecure class (we dropped the constant term in these equations). Using these probability estimates to classify the 2000 test examples of the OMIB problem leads to an error rate of 10.45%.

Obviously, in the OMIB data base, Pu and Qu are far from having Gaussian distributions in each class, as one may induce from Fig. 3.6. This explains the very bad performance of the above naive Bayes classifier. We could relax the Gaussian hypothesis by using the one-dimensional histograms of Fig. 3.6 to approximate class-conditional probability distributions of Pu and Qu in a non-parametric way. Doing so with the OMIB data base yields indeed a smaller test set error rate (7.2%). However, though significantly better than the previous one it is still fairly large and we can conclude that the attributes are not independently distributed in each class which is indeed obvious from the scatter plot given in Fig. 1.2.

Thus, our example illustrates some of the pitfalls of blindly making assumptions such as Gaussian structure of probability distributions and independence hypotheses.

3.6 UNSUPERVISED LEARNING

Below we will describe three types of unsupervised learning methods, representative of a large number of algorithms. The first type of methods (hierarchical clustering) are mainly useful in the context of correlation analysis of numerical attributes. The second type (iterative methods) are interesting in order to group objects, e.g. power system security scenarios, into a small number of similar categories. Finally, the third class of methods (mixture distribution fitting) are interesting because they allow one to decompose a probability distribution into a reduced number of components and to estimate the probability of each object to belong to these latter.

3.6.1 *Hierarchical agglomerative clustering*

Hierarchical clustering aims at defining a sequence of clusterings into K groups, for $K \in [1 \dots N]$, so that clusters form a nested sequence, i.e. such that objects which belong to a same cluster at step K remain in the same cluster at step $K-1$.

The top down or divisive approach consists of generating this sequence in the order of increasing values of K. In the bottom up or agglomerative approach, objects are progressively merged in a step-wise fashion. We briefly describe and illustrate the latter.

The agglomerative algorithm starts with the initial set of N objects, considered as N singleton clusters. At each step it proceeds by identifying the two most similar clusters and merging them to form a single new cluster. This process continues until all objects have been merged together in a single cluster. Cluster similarity may be defined in various ways, for example combining object similarities as follows

$$SIM_{\min}(\text{Cluster}_i, \text{Cluster}_j) = \min_{o_i \in \text{Cluster}_i; o_j \in \text{Cluster}_j} SIM(o_i, o_j), \quad (3.37)$$

$$SIM_{\max}(\text{Cluster}_i, \text{Cluster}_j) = \max_{o_i \in \text{Cluster}_i; o_j \in \text{Cluster}_j} SIM(o_i, o_j). \quad (3.38)$$

The resulting hierarchical clustering may be represented by a dendrogram which shows graphically the hierarchical groupings of objects along with the cluster similarities. This is particularly interesting for the analysis of attribute similarities or when the number of objects to cluster is small.

Illustration. An example dendrogram is represented in Fig. 3.7. It was built for the hierarchical clustering of attributes for the transient stability example of §5.3 relative to the Hydro-Québec power system.

For this illustration, we have considered a selection of 14 power flow and 3 power generation attributes chosen among 67 candidate attributes. The similarity among the attributes was defined as their correlation coefficient (see eqn. (2.14)) which was estimated for each pair of attributes on the basis of the 12497 operating states of the data base.

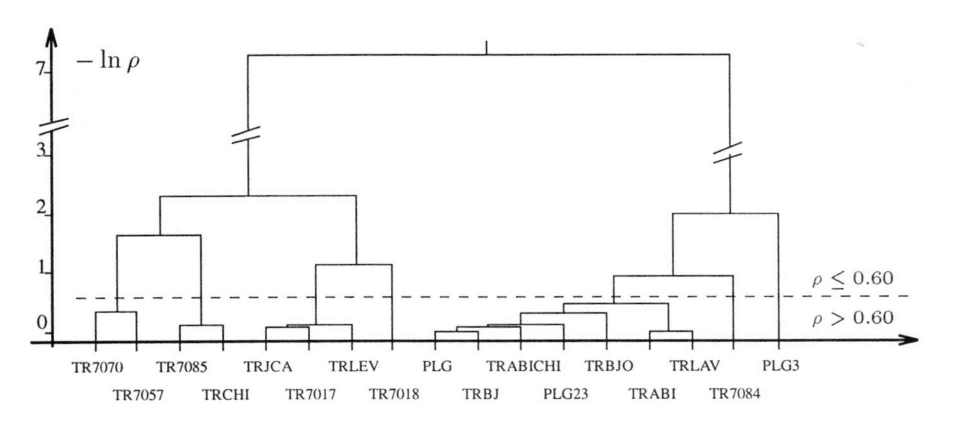

Figure 3.7. Hierarchical attribute clustering example

The similarity of two subsets of attributes S_1 and S_2 was defined as the minimum similarity of pairs of attributes of the two subsets, using eqn. (3.37). To improve the graphical rendering we have used the logarithm of the inverse of this similarity measure to draw the dendrogram in Fig. 3.7. Thus the vertical position of the line merging to clusters represents the following quantity

$$\text{Distance}(S_1, S_2) \triangleq -\ln \left(\min_{a_1 \in S1; a_2 \in S_2} |\hat{\rho}(a_1, a_2)| \right). \tag{3.39}$$

It is interesting to observe from the dendrogram that PLG, TRBJ, TRABICHI, PLG23, TRBJO, TRABI and TRLAV form a rather homogeneous group of similar attributes, the correlation being at least equal to 0.60 for each pair. Actually, a closer look at these attributes shows that they all correspond either to generations of the LaGrande power plants or to North to South power flows in James's Bay corridor (see Fig. 5.6).

Similarly, the group of attributes TRJCA, TR7017 and TRLEV are related to lines within the Québec to Montréal corridor, which are shared by the North to South and the West to East corridors of the Hydro-Québec system.

3.6.2 Algorithms of dynamic clusters

In this section we will briefly describe the ISODATA and the K-means algorithms which are useful for clustering a large number of objects into a small number of groups.

Given a set of objects, and a number K fixed a priori by the user, these procedures determine a set of K clusters, so as to represent most effectively the prior distribution in the attribute space $p(\boldsymbol{a})$ by the cluster prototypes or centers.

In these methods, a cluster is defined by its *prototype* and its members are the learning states which are most *similar* to the cluster prototype. The iterative algorithms stop as soon as a stable partition of the data has been found.

In the basic algorithm, a prototype is defined as the mean attribute vector of a cluster and the similarity is defined as the Euclidean distance. This leads to the basic ISODATA and K-means algorithms searching for clusters minimizing the following quadratic quantization error criterion

$$J_e = \sum_{i=1,K} J_i, \tag{3.40}$$

where J_i denotes the quantization error of the cluster i, defined by

$$J_i \triangleq \sum_{o \in LS; o \in \text{Cluster}_i} ||\boldsymbol{a}(o) - \overline{\boldsymbol{a}}_i||^2, \tag{3.41}$$

$\overline{\boldsymbol{a}}_i$ denoting the center or prototype of the i-th cluster.

This criterion is clearly sensitive to the normalization of the attributes, and thus the clusters found may strongly depend on the normalization. In order to achieve invariance, one should therefore transform the attributes using one of the techniques described in §§6.2.1 or 6.2.3. However, this may also be detrimental in some situations. Thus the definition of a clustering criterion is essentially a problem solved in an empirical, pragmatic trial and error fashion.

The so-called *dynamic clustering algorithm* is a generalization of the ISODATA method, which allows us to use a general class of kernels for representing prototypes and employs a more general similarity based optimality criterion [DK82].

ISODATA. In the ISODATA algorithm, the cluster centers are adapted iteratively in the following *batch* fashion.

1. Choose the initial cluster prototypes randomly or on the basis of prior information.
2. Classify all the learning states by allocating them to the closest cluster.
3. Recompute the K prototypes on the basis of their corresponding learning states.
4. If at least one cluster prototype has changed, return to step 2, otherwise stop.

K-means. This quite similar approach starts with the definition of the initial clusters as given sets of objects, and operates schematically in the following sequential *on-line* fashion.

1. Start with a random partition of the data into K clusters, and compute the corresponding cluster centers as the means of each cluster's objects' attribute vectors.

2. Select the next candidate object o from the learning set, and let i be its current cluster.

3. (a) If o is in a single object cluster then this remains unchanged.

   (b) Otherwise find the cluster j which results in a minimum overall quantization error J_e, if object o is moved from cluster i to cluster j. If $i \neq j$ move the object and adapt both cluster centers.

4. If J_e has remained unchanged during a complete pass through the learning set then stop, otherwise return to step 2.

This latter approach has the advantage of being sequential and thus may be applied in real time, in order to adapt the clustering to new incoming objects. Its main disadvantage, with respect to the ISODATA batch algorithm is its higher susceptibility of being trapped in local minima [DH73].

The initial random partition of the data may be obtained as follows : (i) choose K random objects in the $\boldsymbol{LS}$ as initial cluster centers; (ii) classify all objects into their closest cluster.

Determining the right number of clusters. In practice the number of clusters is often unknown and must also be determined on the basis of the data. The classical approach to this problem consists of applying either of the above algorithms repeatedly with a growing (or decreasing) number of clusters K.

In practice, for each value of K a performance measure is computed for the corresponding clusters obtained. For example, in the above mean square error framework, the overall quantization error $J_e(K)$ could be used for this purpose. The $J_e(K)$ criterion decreases towards zero when K increases and an appropriate number of clusters may be selected by detecting the value of K corresponding to a "knee" in the $J_e(K)$ curve, above which J_e decreases much more slowly.

3.6.3 Mixture distribution fitting

To conclude our brief introduction to unsupervised learning let us mention an important family of methods which approach this problem in a probability distribution fitting paradigm.

In this framework one considers the hypothesis that the learning sample was generated by a probability distribution $p(\boldsymbol{a})$ which is supposed to be a mixture of K elementary probability densities corresponding to the elementary underlying classes under investigation.

To illustrate this idea, we will merely describe the basic principle of the recent *AutoClass* algorithm, but many other methods have been proposed within this framework; for further information the interested reader may refer to [DH73, Han81, DK82].

AutoClass. AutoClass is based on a Bayesian approach to clustering, proposed by Cheeseman [CSKS88]. Its main advantages are its ability to determine automatically the most likely number of clusters and to handle both numerical and symbolic attributes, including hierarchically structured ones. The main assumption of AutoClass is that the attributes values are independent in a given cluster. Thus each cluster is described by a product of elementary attribute distributions. Real-valued attributes are modeled by one-dimensional Gaussian distributions and discrete attributes by probability tables.

The AutoClass approach is based on the Bayesian theory of finite mixtures. Each learning state is assumed to be drawn from one of K mutually exclusive and exhaustive classes, described by a probability distribution as indicated above :

$$p(\boldsymbol{a}|M) = \sum_{j=1,K} p_j p(\boldsymbol{a}|c_j, M_j), \tag{3.42}$$

where M denotes the model hypothesis which is composed of the vector of class probabilities p_j and one set of model parameters M_j for each class c_j.

For a given choice of the model parameters M, each observation $\boldsymbol{a}$ will have a probability of belonging to each class computed by

$$p(c(o) = c_i|\boldsymbol{a}(o), M) = \frac{p_i p(\boldsymbol{a}(o)|c(o) = c_i, M_i)}{p(\boldsymbol{a}(o)|M)}. \tag{3.43}$$

To learn the model parameters and its order K, the joint probability density of the $\boldsymbol{LS}$ under the model assumption and independence hypothesis is computed

$$p(\boldsymbol{LS}|M) = \prod_{o \in \boldsymbol{LS}} p(\boldsymbol{a}(o)|M). \tag{3.44}$$

From this, the posterior distribution of the model parameters may be computed, under the hypothesis of known order K by

$$p(M|\boldsymbol{LS}, K) = \frac{p(M|K)p(\boldsymbol{LS}|M, K)}{p(\boldsymbol{LS}|K)}, \tag{3.45}$$

where $p(\boldsymbol{LS}|K)$ is the normalizing constant obtained by

$$p(\boldsymbol{LS}|K) = \int_{\mathcal{M}_K} p(M_K|K)p(\boldsymbol{LS}|M_K, K)dM_K, \tag{3.46}$$

where M_K denotes the parameter choice for a model of order K, and $\mathcal{M}_K$ the space of possible such models.

The posterior distribution of the number of classes is then obtained by

$$p(K|\boldsymbol{LS}) = \frac{p(K)p(\boldsymbol{LS}|K)}{p(\boldsymbol{LS})}. \tag{3.47}$$

The optimal order is the one maximizing the above probability.

It is important to notice that there are two prior distributions, $p(K)$ and $p(M_K|K)$, which must be filled in the above reasoning in order to define the algorithm.

In particular, we may for example assume that the prior distributions of the model complexity are uniform and that the model parameters are conditionally distributed uniformly, i.e. $p(M_K|K)$ is uniform in an a priori defined parameter interval. In this case, the prior probability of a particular choice of parameters $p(M, K)$ will automatically decrease when the number of parameters increases. And this decrease in prior probability will trade off the increased model fit $p(\boldsymbol{LS}|M, K)$ in eqn. (3.45) and prevent over-fitting. Of course, the algorithm may also take into account the user's prior beliefs about model complexity and parameters.

Thus, the apparently inconsequent hypothesis of a *conditional* uniform prior model probability given its complexity, leads to the cost complexity tradeoff. This should be compared with the maximum likelihood strategy, which is equivalent to assuming a priori that all models are equally likely, *independently* of their complexity.

A very similar reasoning leads to the Bayesian justification of the tree quality measure explained in §5.3 [Weh94].

3.7 FURTHER READING

In this chapter we have screened a representative sample of statistical automatic learning methods. However, our presentation is far from being complete in terms of both the methods discussed and the amount of technical information provided. Thus, we encourage the interested reader to refer to the very reach literature on the subject, a sample of which is given below.

References

[DK82] P. A. Devijver and J. Kittler, *Pattern recognition : A statistical approach*, Prentice-Hall International, 1982.

[DH73] R. O. Duda and P. E. Hart, *Pattern classification and scene analysis*, John Wiley and Sons, 1973.

[Han81] D. J. Hand, *Discrimination and classification*, John Wiley and Sons, 1981.

[MST94] D. Michie, D.J. Spiegelhalter, and C.C. Taylor (eds.), *Machine learning, neural and statistical classification*, Ellis Horwood, 1994, Final rep. of ESPRIT project 5170 - StatLog.

[Jen96] F. V. Jensen, *An introduction to Bayesian networks*, UCL Press, 1996.

4 ARTIFICIAL NEURAL NETWORKS

4.1 OVERVIEW

While the learning systems based on artificial neural networks became popular only in the early eighties, they have a much longer research history and some of these methods have evolved towards quite mature techniques.

The early work on neural networks was mainly motivated by the study and modeling of human learning abilities. In particular, the neural network approach aimed at reproducing these latter by developing a low-level model reflecting the biological structure of the brain. Such artificial neural models date back to the 1940's, with the work by McCulloch and Pitts [MP43] on modeling the brain neuron behavior.

The second wave of the research reached its peak in the early sixties with the perceptron learning theorem of Rosenblatt [Ros63], which proved that the simple on-line error-correcting learning algorithm proposed would converge in a finite number of steps if the patterns are linearly separable. However, the negative results of Minsky and Papert [MP69], concerning the perceptron's representation capabilities, brought a firm stop to the enthusiasm.

Finally, the present wave has started from the conjunction of the rapid increase in available computing power in the early 1980's, the theoretical work of Hopfield, and the improvements of multilayer perceptrons culminating with the (re)publication of the back-propagation algorithm by Rumelhart, Hinton and Williams [RHW86].

Thus, since the mid 1980's, an almost exponentially growing amount of theoretical and practical work has been published, leading to the creation of new journals and conferences, and several textbooks. Since it would be very difficult to give a reasonably representative account of this work, we will restrict our focus to three methods : multilayer perceptrons, radial basis functions (which are closely related), and Kohonen feature maps. At the end of this chapter we will provide some guidelines for further reading on this subject matter.

As we mentioned in Chapter 1, multilayer perceptrons are essentially a nonlinear regression technique, closely related to the non-parametric statistical regression techniques such as projection pursuit. However, perceptrons (their ancestor) were designed for classification problems and multilayer perceptrons were proposed to overcome perceptrons' limitations in this context. Thus, keeping in mind that multilayer perceptrons are most useful in the context of regression problems, below we will follow the historical path in our description, by first considering classification problems.

4.2 SINGLE LAYER PERCEPTRONS

We start by describing the "linear threshold unit" model of a neuron and its learning algorithm, then we generalize it to soft thresholds and the minimum square error learning algorithm.

4.2.1 Linear threshold units

Description. The linear threshold unit (LTU) is a hyper-plane model similar to the linear discriminants of §3.2.1 (see Fig. 4.1). It is defined by a discrimination function[1]

$$g(\boldsymbol{a}(o)) \triangleq \text{sign}\left\{w_0 + \boldsymbol{w}^T \boldsymbol{a}(o)\right\} \tag{4.1}$$

which assigns a value of -1 or $+1$ depending on which side of the hyper-plane corresponding to its weight vector, the attribute vector $\boldsymbol{a}(o)$ is located. This model may be used in order to solve a two-class classification problem either for real-valued attributes or boolean ones coded as binary -1/+1 indicators.

Supposing that the learning set classification information has been encoded in the above binary fashion, i.e. $c(o) \in \{-1, +1\}$, the ideal objective of the perceptron learning algorithm is to reproduce the learning set classification perfectly, or equivalently to choose an extended weight vector $\boldsymbol{w}'^T = (w_0, \boldsymbol{w}^T)$ such that

$$\sum_{o \in LS} \left[g(\boldsymbol{a}(o)) - c(o)\right]^2 = 0. \tag{4.2}$$

Learning problems for which this equation has a solution are called linearly separable.

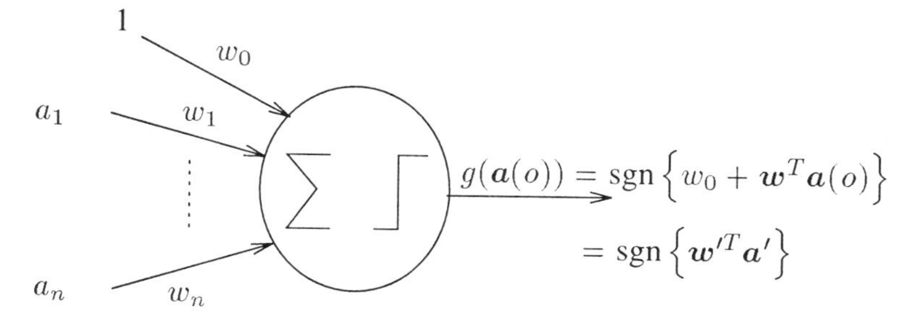

Figure 4.1. Basic linear threshold unit

Table 4.1. Perceptron learning algorithm

1. *Consider the objects of the learning set in a cyclic or random sequence.*
2. *Let o be the current object,* c(o) *its class and* $\boldsymbol{a}$(o) *its attribute vector.*
3. *Adjust the weight by using the following correction rule,*

$$\boldsymbol{w}'^{\text{new}} = \boldsymbol{w}'^{\text{old}} + \eta\,(c(o) - g(\boldsymbol{a}(o)))\,\boldsymbol{a}'(o). \tag{4.3}$$

Learning algorithm. The so-called perceptron learning algorithm for LTUs is given in Table 4.1.

It is a sequential "on-line" method considering successive passes through the learning states, and adjusting the weights at each step so as to improve the classification of the current object, if necessary. Notice that eqn. (4.3) implies that the correction is equal to zero for objects which are already classified correctly. On the other hand, for an incorrectly classified object, the weight vector is corrected so as to bring the output closer to the desired output value $c(o)$.

The parameter η denotes the *learning rate* of the algorithm, and various strategies have been proposed to choose its value. It may be shown that if the learning set is separable, then the fixed learning rate perceptron learning rule converges to a solution in a finite number of steps, but the speed of convergence may depend on the precise value of η. In addition, if the learning set is not separable, then the algorithm will never stop changing the weight values. Thus, one of the techniques used to ensure convergence consists of using a decreasing sequence of learning rate values $\eta_k \to 0$.

Discussion. The LTU may be generalized to a single layer of LTUs, in order to be able to learn a boolean vector or a binary coded integer output function, for example to solve multi-class classification problems.

However, these simple models are unable to represent an as simple function as the two-dimensional logical XOR (exclusive OR) operator, or the general n-dimensional parity function.

The solution to these types of problems calls for multilayer structures with nonlinear but differentiable input/output relations, to allow the use of the error back-propagation learning algorithm.

Let us first consider *soft* threshold units which provide the elementary brick to build up such general powerful multilayer models.

4.2.2 Soft threshold units and minimum square error learning

The soft threshold unit is a slight modification of the perceptron, where the discontinuous sign function is replaced by a smooth (nonlinear) activation function.

The input/output function $g(\boldsymbol{a})$ of such a device is computed by

$$g(\boldsymbol{a}(o)) \triangleq f(w_0 + \boldsymbol{w}^T \boldsymbol{a}(o)) = f(\boldsymbol{w}'^T \boldsymbol{a}'(o)) \tag{4.4}$$

where the *activation* function $f(\cdot)$ is assumed to be differentiable. Classical examples of activation functions are the sigmoid

$$\text{sigmoid}(x) = \frac{1}{1+\exp(-x)}, \tag{4.5}$$

and the hyperbolic tangent

$$\tanh(x) = \frac{\exp(x) - \exp(-x)}{\exp(x) + \exp(-x)}, \tag{4.6}$$

but other types of general (linear or nonlinear) smooth functions may also be considered.

Considering output values varying continuously between -1 and $+1$, and the possibility of non-separable problems, we now reformulate the learning objective as the definition of a weight vector $\boldsymbol{w}' = (w_0, \boldsymbol{w})$ minimizing the mean square error (SE)

$$SE(\boldsymbol{w}') \triangleq N^{-1} \sum_{o \in LS} |g(\boldsymbol{a}(o)) - c(o)|^2 . \tag{4.7}$$

The gradient of the SE with respect to the augmented weight vector $\boldsymbol{w}'$ is computed by

$$\nabla_{\boldsymbol{w}'} SE = 2N^{-1} \sum_{o \in LS} (g(\boldsymbol{a}(o)) - c(o))\, f'(\boldsymbol{w}'^T \boldsymbol{a}'(o)) \boldsymbol{a}'(o), \tag{4.8}$$

where $f'(\cdot)$ denotes the derivative of the activation function $f(\cdot)$ and $\nabla_{\boldsymbol{w}'}$ is the gradient operator in the weight space.

Thus, using a *fixed* step gradient descent approach for minimizing the mean square error, in a sequential object by object correction setting, would consist of using the following weight update rule

$$\begin{aligned} \boldsymbol{w}'^{new} &= \boldsymbol{w}'^{old} - \eta \nabla_{\boldsymbol{w}'} SE(o) \quad (4.9) \\ &= \boldsymbol{w}'^{old} + \eta \left[c(o) - g(\boldsymbol{a}(o))\right] f'(\boldsymbol{w}'^{T} \boldsymbol{a}'(o)) \boldsymbol{a}'(o), \quad (4.10) \end{aligned}$$

where $SE(o)$ denotes the contribution of object o in eqn. (4.7).

This is analog to the perceptron learning rule of Table 4.1, where the *learning rate* η is adapted proportionally to the derivative f' of the activation function.

An alternative, *batch* learning approach consists of computing the full gradient (4.8) of the SE with respect to the complete learning set before correcting the weight vector.

A further improvement would then consist of using a *variable* step gradient descent method, for example the *steepest descent* approach. This consists of using a line search so as to determine at each stage the step (η) in the (opposite) gradient direction resulting in a maximal decrease of the SE criterion. Other more sophisticated numerical optimization techniques may be thought of (see §4.3.5). Their main property is that each iteration leads to a decrease of the error function; they are thus guaranteed to a (local) minimum of the error function.

However, one of the remaining problems concerns the existence of several local minima of the SE criterion. The particular one to which the gradient type search techniques will converge will mainly depend on the initial weight values. Thus, in order to increase the probability of finding a global minimum, one should repeat the search procedure for various randomized initial weight vectors. Another suggestion has been to apply heuristic global optimization techniques such as the simulated annealing method or the genetic algorithms discussed in §6.3.

Finally, in the context of classification problems the SE objective function minimization does not necessarily lead to a minimum number of misclassification errors, neither in the learning set, nor - a fortiori - in an independent test set.

4.3 MULTILAYER PERCEPTRONS

Multilayer perceptrons (MLPs) are composed of several layers of soft threshold units, interconnected so as to enable the approximation of arbitrary nonlinear input/output relationships.

Back-propagation is the extension of the gradient descent learning algorithm to such multilayer structures. Its name stems from the fact that this algorithm computes the gradient of the MLP output error with respect to the weights by propagating information backwards through the MLP.

Notice that many authors refer to the back-propagation algorithm as the combination of computing the gradient in the weight space and the adaptation of the weights, with a fixed step gradient descent approach. Below, we refer to back-propagation as the sole first step, i.e. the computation of the gradient of neuron activations in the weight space.

The popularity of the back-propagation algorithm in the ANN community stems from the fact that it is a local method, which allows to adapt weights feeding a neuron on the sole basis of information provided by its immediate neighbors. This, together with memory-less learning (learning in an on-line scheme, without storing the whole learning set) was considered in the early eighties as a requirement for *biological plausibility* (this is not true anymore today).

The other, to us more important feature of back-propagation is that it is an *efficient* algorithm to compute the derivatives of the output of the network with respect to its weights. As we will see, this algorithm's complexity is linear in the number of weights, thus is can be applied to very large neural networks containing a very large number of connections. It is thus also exploited in more sophisticated (non-local and batch) optimization algorithms mostly used nowadays for MLP training (see §4.3.5).

4.3.1 Back-propagation for feed-forward structures

The basic idea of the algorithm is to compute the derivatives of the error function in a layer by layer fashion, starting with weights feeding the output layer and ending with the weights feeding the first hidden layer of neurons. In order to highlight fully the scope of the algorithm we will formulate in a very general manner. Then, we will apply it to the single hidden layer structure of §1.2.4.

Let us consider the general feed-forward structure suggested in Fig. 4.2, where the neurons are sequentially ordered from 1 to K. In this structure a neuron j receives a net input $n_j(o)$

$$n_j(o) \triangleq \sum_{i=1,j-1} w_{i,j} x_i(o), \tag{4.11}$$

where $w_{i,j}$ denotes the weight of the connection from neuron i to neuron j, and $x_i(o)$ the activation (or output) of neuron i, for object o^2. Further, each neuron j has a differentiable activation (or transfer) function $f_j(\cdot)$ and its output state x_j is computed by

$$x_j(o) \triangleq f_j(n_j(o)). \tag{4.12}$$

Although the classical multi-*layer* perceptron is a particular case of this structure, where some of the weights are constrained to be equal to zero, it is simpler to explain the back-propagation algorithm on the basis of the above general, maximally connected feed-forward structure.

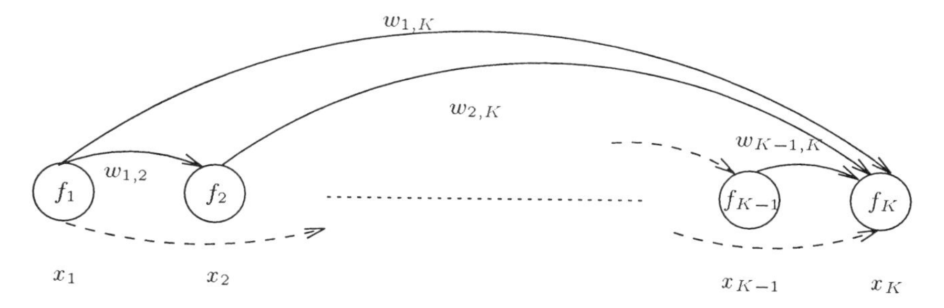

Figure 4.2. General feed-forward network

Further, we consider the following general type of "error" function

$$ERR(w_{i,j}, LS) \triangleq \sum_{o \in LS} h(\boldsymbol{x}(o), \boldsymbol{y}(o)), \tag{4.13}$$

where $h(\cdot, \cdot)$ denotes a differentiable function of the neuron activation vector $\boldsymbol{x} = (x_1, \ldots, x_K)$ and of the desired output vector $\boldsymbol{y} = (y_1, \ldots, y_r)$. The usual SE criterion is a particular case of this "error" function.

The derivatives of the error function with respect to the network weights $w_{i,j}$ are then computed by the following formula

$$\frac{\partial ERR(w_{i,j}, LS)}{\partial w_{i,j}} = \sum_{o \in LS} \sum_{k=1,K} \frac{\partial h(\boldsymbol{x}(o), \boldsymbol{y}(o))}{\partial x_k} \frac{\partial x_k}{\partial w_{i,j}}. \tag{4.14}$$

On the other hand, the essence of the back-propagation algorithm which is suggested graphically in Fig. 4.3, consists of computing the partial derivatives

$$\frac{\partial x_k}{\partial w_{i,j}}, \tag{4.15}$$

by propagating them back from the high order to the low order neurons.

More precisely, these derivatives are obtained by the following equation

$$\frac{\partial x_k}{\partial w_{i,j}} = x_i \delta_j, \tag{4.16}$$

where the values of δ_j are derived by the following backward recursion relations

$$\delta_j = 0, \ \forall j > k; \tag{4.17}$$

$$\delta_j = f'_j(n_j), \ \forall j = k; \tag{4.18}$$

$$\delta_j = f'_j(n_j) \sum_{p=j+1,k} w_{j,p} \delta_p, \ \forall j < k. \tag{4.19}$$

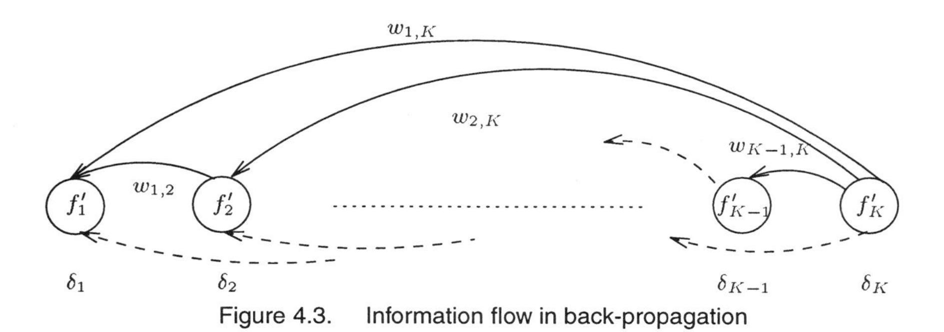

Figure 4.3. Information flow in back-propagation

We omit the proof of the above equations and refer the interested reader to [Hay94] for further details.

These relations can be used in order to compute derivatives of a neuron activation with respect to all weights in a single back-propagation pass (see illustration below).

If there are several (output) neurons whose activation is taken into account in the function $h(\cdot)$, several passes would be required, one for each output neuron. However, due to the linearity of eqns. (4.14) and (4.19), this can be done in a single pass, by feeding backwards the derivatives $\frac{\partial h}{\partial x_k}$ from all output neurons at the same time.

While the above recursion is a simple chain-rule differentiation, the interesting point is that the corresponding error back-propagation algorithm is local and uses the same network structure than the original feed-forward network to propagate information, though in the reverse direction. Another notable fact is the surprising computational efficiency of this algorithm, since *all* the derivatives are obtained with the order of w operations, where w is the total number of weights of the network. Furthermore, the algorithm can be slightly modified to compute the derivatives of some neuron activations (e.g. outputs) with respect to other neuron activations (e.g. inputs).

Finally, the method may be used either in an incremental learning scheme, adapting the weights after each presentation of an input attribute vector, or in the batch approach cumulating derivatives over the full learning set before adapting the weights.

4.3.2 Multilayer structure

In a classical *multilayer* feed-forward network, the first $n+1$ neurons would correspond to the extended input attribute vector $\boldsymbol{a}'$. Thus, their activation would be fixed, for a given object o presented to the network, independently of any weight values by

$$x_j(o) = a_j(o), \forall j = 1, n, \text{ and } w_{i,j} = 0, \forall i < j, \tag{4.20}$$

and

$$x_{n+1}(o) = 1 \text{ and } w_{i,n+1} = 0, \ \forall\ o. \tag{4.21}$$

On the other hand, the last r output values would correspond to the output information of the network,

$$x_{k-r+j}(o) = r_j(o), \forall j = 1, \ldots, r \text{ and } w_{j,i} = 0, \forall i > j. \tag{4.22}$$

Finally, the layers are defined as groups of consecutive neurons receiving information from neurons in the preceding layer and feeding their information to neurons in subsequent layers. Denoting by L_l a layer number l, and L the total number of layers

$$L_l = \{j, j+1, \ldots, j+n_L-1\}, l = 1, \ldots, L \tag{4.23}$$

this corresponds to a set of connectivity constraints

$$w_{i,j} = 0, \forall i, j | i \in L_l, j \notin L_{l+1}. \tag{4.24}$$

In the case of the multilayer structure, the error function would explicitly take into account only neuron activations of the last (output) layer. For example, the standard SE error function is defined by

$$h(\boldsymbol{x}(o), \boldsymbol{y}(o)) = N^{-1} \sum_{i=1,r} |\boldsymbol{x}_{k-r+i}(o) - y_i(o)|^2. \tag{4.25}$$

The overall derivative of the error function is then obtained by sweeping through the learning set, and computing for each object the activation vector $\boldsymbol{x}(o)$ in the feed-forward fashion, and using the back-propagation algorithm to compute the derivatives with respect to each weight in a backward fashion, cumulating the terms corresponding to the components of the activation vector which are explicitly used in the error function, proportionally to the corresponding partial derivative $\frac{\partial h}{\partial x_h}$. All the computations being linear, this may be done in a single pass for all output neurons, and for a given network activation corresponding to a given object.

4.3.3 *Illustration*

Let us illustrate the back-propagation algorithm on our OMIB example of §1.2.2 and the MLP of §1.2.4. We consider the MLP weights of the single hidden layer network of Fig. 1.10 after a random initialization and illustrate how the algorithm is applied in order to compute the derivatives of the activation of the output neuron with respect to some of the weights in the two layers, for a given learning state presented at the input.

Let us for example consider state number 1, described in Table 1.1. It is characterized by Pu= 876.0MW, Qu= -193.7MVAr. The MLP does not operate directly on these values; they are first pre-whitened by subtracting the mean and dividing by their standard deviation, as computed in the learning set.

The mean value of Pu is equal to 1003MW in the learning and its standard deviation is of 169.3MW. Thus the pre-whitened value of Pu of state 1 is $\frac{876-1003}{169.3} = -0.750$.

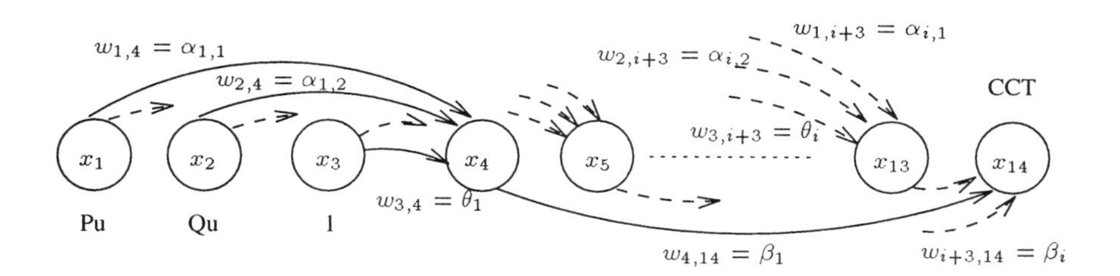

Figure 4.4. MLP for the OMIB example

The mean value of Qu is equal to 156.6MVAr in the learning and its standard deviation is of 480.4MVAr. Thus the pre-whitened value of Qu of state 1 is $\frac{-193.7-156.6}{480.4} = -0.729$.

Figure 4.4 reproduces the MLP of Fig. 1.10 using the notation of §4.3.1 and Fig. 4.2 and showing the correspondence with the notation used in §1.2.4. From left to right, the first two neuron activations x_1 and x_2 correspond to the inputs Pu and Qu; the third one corresponds to a constant input $x_3 = 1$ used for the offset thresholds θ_i; the next 10 neurons ($i = 4, \ldots 13$) correspond to the hidden layer; the last neuron corresponds to the network output, whom we would like to approximate the CCT. Notice that the hidden neurons use the hyperbolic tangent function ($f_4(x), \ldots, f_{13}(x) = \tanh(x)$) whereas the output neuron use the identity function (f_{14} is $f_{14}(x) = x$).

Computation of neuron activations. Let us start by computing the activations in a feed-forward manner of all hidden neurons, assuming that the input neurons are activated with the values of Pu and Qu of state 1.

A hidden neuron i, $i = 1, \ldots, 10$ receives a net input

$$n_{i+3}(\text{state}) = \alpha_{i,1}P_u(\text{state}) + \alpha_{i,2}Q_u(\text{state}) + \theta_i \tag{4.26}$$

and produces output (i.e. activation x_{i+3})

$$out_i(\text{state}) = x_{i+3}(\text{state}) = f_{i+3}(n_{i+3}(\text{state})). \tag{4.27}$$

where the activation function $f_{i+3}(x) = \tanh(x)$ for each hidden neuron.

The initial values of the weights are given in Table 4.2 (all the weights have been initialized with random values in the interval $[-1 \ldots 1]$).

Thus, for example, $\alpha_{1,1} = 0.336204$, $\alpha_{1,2} = -0.168107$ and $\theta_1 = -0.168458$, and we can derive that

$$n_4(\text{state}_1) = 0.336204(-0.75) - 0.168107(-0.729) - 0.168458 = -0.298061$$

and its activation

$$x_4(\text{state}_1) = \tanh(-0.298061) = -0.289537.$$

Table 4.2. Weights and activations of OMIB MLP for state 1

hidden neuron i	$\alpha_{i,1}$ $= w_{1,i+3}$	$\alpha_{i,2}$ $= w_{2,i+3}$	θ_i $= w_{3,i+3}$	n_{i+3}	out_i $= x_{i+3}$	β_i $= w_{i+3,14}$	$\beta_i out_i$
1	0.336204	-0.168107	-0.168458	-0.298061	-0.289537	-0.319319	0.092455
2	0.039182	-0.332096	-0.081034	0.131677	0.130921	-0.213041	-0.027892
3	0.380301	-0.303366	-0.211199	-0.275271	-0.268523	0.246001	-0.066057
4	0.473159	0.187676	0.408981	-0.082704	-0.082516	-0.445895	0.036794
5	0.257133	-0.092062	0.107661	-0.018076	-0.018074	0.468682	-0.008471
6	-0.139672	0.460809	-0.214887	-0.446063	-0.418657	0.131887	-0.055215
7	0.011983	-0.454129	-0.417836	-0.095763	-0.095472	-0.496104	0.047364
8	-0.057184	-0.184123	-0.246296	-0.069182	-0.069072	0.343700	-0.023740
9	-0.167455	-0.356006	-0.236563	0.148557	0.147473	0.137117	0.020221
10	0.004473	-0.228770	-0.045882	0.117537	0.116998	-0.046011	-0.005383

The activations of all the other hidden neurons ($x_{i+3}, i = 2, \ldots, 10$) are computed in a similar fashion. The resulting values are given in the sixth column of Table 4.2. These values are then multiplied by β_i to yield the last column, and summed up to provide the net input of the output neuron

$$n_{14}(\text{state}) = \sum_{i=1,10} \beta_i out_i(\text{state}) = \sum_{i=1,10} \beta_i x_{i+3}(\text{state}) \tag{4.28}$$

The final output provided by the MLP is

$$x_{14}(\text{state}) = \text{CCT}_{\text{MLP}}(\text{state}) = f_{14}(n_{14}(\text{state})) = n_{14}(\text{state}), \tag{4.29}$$

since we have chosen to use a linear output neuron ($f_{14}(x) = x$).

Making the computation, we find that $\text{CCT}_{\text{MLP}}(\text{state}_1) = x_{14} = n_{14} = 0.010075$. The square estimation error for state 1 is thus

$$|\text{CCT}(\text{state}_1) - \text{CCT}_{\text{MLP}}(\text{state}_1)|^2 = |\ 0.236 - 0.010075\ |^2 = 0.051042.$$

Notice that output guessed by the MLP is very far from the actual CCT computed by numerical simulation. This is normal, since we are only at the first step of the learning procedure and the weights were initialized at random.

Computation of derivatives of the output neuron activation. Now, let us consider the computation of the derivatives of the output neuron activation x_{14} (eqn. (4.29)) with respect to some weights, using the back-propagation algorithm according to the formulas (4.16 - 4.19).

Eqn. (4.16) becomes

$$\frac{\partial x_{14}}{\partial w_{i,j}} = x_i \delta_j . \tag{4.30}$$

We start by computing the derivatives with respect to weights $\beta_i = w_{i+3,14}$ feeding the output neuron

$$\frac{\partial x_{14}}{\partial w_{i+3,14}} = x_{i+3}\delta_{14}. \tag{4.31}$$

Since $j = k = 14$, the base case eqn. (4.18) is used to derive $\delta_{14} = f'_{14}(n_{14})$. In our case, $\delta_{14} = f'_{14}(n_{14}) = 1$, since the output layer is linear. Hence,

$$\frac{\partial x_{14}}{\partial w_{i+3,14}} = x_{i+3}\delta_{14} = x_{i+3}. \tag{4.32}$$

Thus, the values of these derivatives of the MLP output w.r.t. β_i are given by the sixth column of Table 4.2.

The next step consists of computing the derivatives of x_{14} with respect to the weights feeding the hidden neurons $(\alpha_{i,1}, \alpha_{i,2}, \theta_i)$. For example, let us consider the derivative with respect to $\alpha_{1,1} = w_{1,4}$, a weight feeding the first hidden neuron (neuron 4, in our notation) . Eqn. (4.30) now becomes

$$\frac{\partial x_{14}}{\partial w_{1,4}} = x_1\delta_4. \tag{4.33}$$

Notice that $x_1 = Pu$, and since $j = 4 < k = 14$ we must apply the recursion eqn. (4.19) to compute δ_4. Thus

$$\delta_4 = f'_4(n_4) \sum_{p=5,14} w_{4,p}\delta_p, \tag{4.34}$$

where the sum is carried out over all connections departing from neuron 4. In the case of our MLP, there is only one single such connection (the one feeding the output neuron 14). Hence

$$\delta_4 = f'_4(n_4) w_{4,14}\delta_{14}, \tag{4.35}$$

and finally

$$\frac{\partial x_{14}}{\partial w_{1,4}} = x_1 f'_4(n_4) w_{4,14}\delta_{14} = Pu(1 - \tanh^2(n_4))\beta_1\delta_{14} \tag{4.36}$$

where we have exploited the identity $\tanh'(\cdot) = 1 - \tanh^2(\cdot)$.

Substituting in eqn. (4.36) the values of $Pu = -0.75$, $n_4 = -0.298061$, $\beta_1 = -0.319319$ (see Table 4.2) and $\delta_{14} = 1$ yields

$$\frac{\partial x_{14}}{\partial w_{1,4}} = -0.75(0.916168)(-0.319319)1 = 0.219412. \tag{4.37}$$

The derivatives of the output neuron activation with respect to the remaining weights feeding neuron 4 ($\alpha_{1,2} = w_{2,4}$ and $\theta_1 = w_{3,4}$) follows a similar pattern

$$\frac{\partial x_{14}}{\partial w_{2,4}} = x_2 f_4'(n_4) w_{4,14} \delta_{14} \tag{4.38}$$

and

$$\frac{\partial x_{14}}{\partial w_{3,4}} = x_3 f_4'(n_4) w_{4,14} \delta_{14} \tag{4.39}$$

Thus, the computation may be decomposed into two steps, in order to avoid recomputing the common factor $\delta_4 = f_4'(n_4) w_{4,14} \delta_{14}$.

1. Back-propagation of the values of δ_p : the values are multiplied by the weights which are crossed and the derivative of the neuron activation functions (here by $w_{4,14} = \beta_1$ and $f_4'(n_4)$); they are stored locally at each neuron.
2. For each weight $w_{i,j}$, computation of the derivative as the product of the activation x_i of its feeding neuron and the value δ_j stored at its output neuron.

It is clear that the overall number of operations will correspond to one feed-forward pass (computation of neuron activations $x_i = f_i(n_i)$ and their derivatives $f_i'(n_i)$), one feed-backward pass (computation of the values of δ_p for each neuron), and one pass through the weights (to compute the products $x_i \delta_j$). The computational complexity of all three operations is linear in the number of weights.

In our example, the feed-forward pass would actually require 50 multiplications and additions and 10 computations of the hyperbolic tangent function. Back-propagation would require another 40 multiplications and additions.

Weight updating. In the fixed step on-line scheme, we would adapt the weights in a state by state fashion. The derivative of the square error for state 1 with respect to any weight is obtained by

$$\frac{\partial\,|\mathrm{CCT(state)} - \mathrm{CCT_{MLP}(state)}|^2}{\partial w_{i,j}} = 2(\mathrm{CCT_{MLP}(state)} - \mathrm{CCT(state)}) \frac{\partial x_{14}}{\partial w_{i,j}}. \tag{4.40}$$

Thus, we would make the following corrections in order to reduce the output square error

$$\Delta w_{i,j} = -\eta 2(\mathrm{CCT_{MLP}(state)} - \mathrm{CCT(state)}) \frac{\partial x_{14}}{\partial w_{i,j}}. \tag{4.41}$$

In the batch mode, the error function used is the mean square error:

$$SE = N^{-1} \sum_{\mathrm{state} \in LS} |\,\mathrm{CCT(state)} - \mathrm{CCT_{MLP}(state)}\,|^2 . \tag{4.42}$$

To compute the derivative of this sum with respect to each weight, we need to carry out the above computations for each learning state and cumulate the resulting values, by weighting them by the error value for every object presented. Actually, this factor can be fed into the computation at the first step, when computing δ_{14}. In our example, the learning set comprises 3000 states, thus the computation of the SE gradient would require 30,000 computations of the hyperbolic tangent function and 300,000 multiplications and additions. Assuming that computing the hyperbolic tangent is about four multiplications and additions, the overall number of multiplications and additions is about 420,000, which is not so large giving present day computing power.

4.3.4 Other objective functions

In our presentation of the back-propagation gradient computation algorithm, we have insisted on its generality, showing that it is able to handle any kind of feed-forward network structures and may be adapted to general activation and objective functions. Below, in §4.3.5, we will discuss some alternative schemes for exploiting these derivatives in order to optimize the objective function in an efficient way. Here we comment on some frequently used alternative network optimization criteria.

Weight decay. Most of the objective functions which have been used in practice derive directly from the standard minimum SE criterion. They take the following general form

$$SE(w_{i,j}, LS) = N^{-1} \left(\sum_{o \in LS} \sum_{i=1,r} |\boldsymbol{x}_{k-r+i}(o) - y_i(o)|^2 + G(||w_{i,j}||^2) \right), \quad (4.43)$$

where $G(||w_{i,j}||^2)$ denotes a generic "regularisation" term, which aims at improving for the "smoothness" of the input/output mapping. The purpose of the regularisation term is to avoid high frequency components in the input/output mapping so as to reduce over-fitting problems. The approach is also called weight decay, because in practice the function $G(\cdot)$ penalizes large weight values.

In many circumstances, using this kind of approach may improve the generalization capabilities of the network with respect to unseen objects, particularly when the number of parameters becomes large with respect to the size N of the learning set.

Entropy based criteria. Other types of fitting criteria have been derived from the logarithmic entropy function. These are interesting alternatives in the case where the output information corresponds to conditional class-probabilities [RL91]. In this case, we assume that the output neurons correspond to the classes, and, ideally, the output vector would be equal to the vector of conditional class probabilities $\boldsymbol{p}(\boldsymbol{a})$ corresponding to the input attribute vector.

For example, the total residual entropy of the $\boldsymbol{LS}$ classification given the network weights may be defined by

$$NH_{C|w_{i,j}}(LS) \triangleq - \sum_{o \in LS} \log P(c(o)|w_{i,j}, \boldsymbol{a}(o)). \tag{4.44}$$

Here $P(c(o)|w_{i,j}, \boldsymbol{a}(o))$ denotes the activation of the output neuron corresponding to the class $c(o)$ for each object, interpreted as the conditional probability predicted by the neural network model to observe the object's class.

Various theoretical *minimum description length* or *maximum a posteriori probability* interpretations may be derived for this criterion by adding a term which would take into account the encoding length of the weights (complexity) or their prior probabilities. Notice that this criterion is similar to the one which is used to define the quality of decision trees in §5.3.2.

4.3.5 More efficient network optimization algorithms

The most obvious and simple way of using the back-propagation algorithm to optimize the neural network fit to the learning set, is to use the fixed step gradient descent algorithm, which is classically referred to as *the* error back-propagation algorithm [HKP91]. Unfortunately, this approach, already very slow in convergence in the single layer perceptron case, is even much slower in the case of multiple nonlinear layers. In practice, the computing times become rapidly prohibitive as soon as the number of weights and learning states increase. Furthermore, this algorithm requires careful tuning of the step size η used.

In the literature, a number of alternative algorithms have been proposed to speed up the convergence. The earliest methods (which consist basically of adding a heuristic "momentum" term to the gradient) present the advantage of preserving the locality of the weight update rule of the back-propagation algorithm. Unfortunately, these ad hoc methods require, in general, a tedious manual tuning of their parameters, which for large scale problems may become very time consuming.

More recently, a certain number of researchers have proposed the use of some of the classical unconstrained optimization algorithms available from the optimization literature [Wat87, Par87, FD92]. The very important improvement in efficiency obtained with respect to the standard steepest descent algorithm and the fact that no user defined parameters must be tuned make this type of approach very attractive.

Discussing the broad topic of nonlinear function optimization is not our goal here. Rather, we will briefly describe the particular method which we have been using in most of our simulations in the context of power system security problems. This is the "Broyden-Fletcher-Goldfarb-Shanno" (BFGS) quasi-Newton algorithm, already mentioned in Chapter 1.

Basic iterative optimization scheme. The basic scheme of the iterative optimization methods consists of defining at each step of the process a search direction $\boldsymbol{s}$ in the weight space, and searching for the minimum of the error function in this direction. This is a scalar optimization problem

$$\min_{\lambda} ERR(\boldsymbol{w} + \lambda \boldsymbol{s}), \tag{4.45}$$

where $\boldsymbol{w}$ denotes the weight vector.

The steepest descent method consists of moving in the direction opposite to the gradient

$$\boldsymbol{s} = -\nabla_{w_{i,j}} ERR(\boldsymbol{w}). \tag{4.46}$$

This leads to a zigzag optimization path, which most often converges very slowly.

Quasi-Newton optimization. A better approach, at least nearby the solution, would be provided by a Newton-like method consisting of computing the search direction by

$$\boldsymbol{s} = -(\nabla^2 ERR(\boldsymbol{w}))^{-1} \nabla ERR(\boldsymbol{w}). \tag{4.47}$$

Unfortunately, this approach may be inefficient due to the high cost of computing the $\frac{w(w+1)}{2}$ terms of the inverse Hessian matrix $(\nabla^2 ERR(\boldsymbol{w}))^{-1}$.

Thus, the basic idea of the quasi-Newton family of methods consists of building up iteratively an approximation of the Hessian matrix inverse from repeated computations of the gradient.

More precisely, the BFGS variant which we have used, is based on the following update scheme at step k [FD92]

$$\boldsymbol{s}^k = -H^k \nabla ERR(\boldsymbol{w})^k, \tag{4.48}$$

and

$$H^{k+1} = H^k + \left(1 + \frac{\boldsymbol{\delta}_\nabla^T H^k \boldsymbol{\delta}_\nabla}{\boldsymbol{\delta}^T \boldsymbol{\delta}_\nabla}\right) \frac{\boldsymbol{\delta}\boldsymbol{\delta}^T}{\boldsymbol{\delta}^T \boldsymbol{\delta}_\nabla} - \left(\frac{\boldsymbol{\delta}\boldsymbol{\delta}_\nabla^T H^k + H^k \boldsymbol{\delta}_\nabla \boldsymbol{\delta}^T}{\boldsymbol{\delta}^T \boldsymbol{\delta}_\nabla}\right), \tag{4.49}$$

where $\boldsymbol{\delta}$ denotes the change of the weight vector at step k as determined by the optimal search in the direction $\boldsymbol{s}^k$ and $\boldsymbol{\delta}_\nabla$ denotes the change in the gradient direction from step k to step $k+1$. The method starts with an initial guess H^0 of the inverse Hessian matrix which is generally taken as the identity matrix. It follows from eqn. (4.47) that in this particular case, the first iteration of steepest descent and BFGS are identical.

As we have observed in practice, the use of this method allows us to considerably reduce the computational burden of the neural network learning, without requiring any manual tuning of parameters.

These quasi-Newton methods, together with the conjugate gradient methods, are the most prevailing efficient techniques used in the context of feed-forward network learning. Nevertheless, they still remain iterative in essence. In particular for real life, medium to large scale problems, they may still require a large number of rather lengthy iterations, without guaranteeing global optimization.

4.3.6 Theoretical properties

In our intuitive introduction to multilayer perceptrons we mentioned that single hidden layer perceptrons are sufficiently powerful to represent continuous input/output relationships. In fact, it has been shown that if a large enough number of hidden neurons is used, they can arbitrarily well approximate any continuous function on a compact subset of R^n [Cyb89, HSW89]. More recently Barron has studied the bias/variance tradeoff in such networks [Bar93], which is a much more practically relevant topic. Below we will merely explain the type of results he has obtained, and how they may be exploited in practice.

Barron shows that in single hidden layer sigmoidal perceptrons bias is upper bounded as follows

$$\text{Bias}^2 \leq \frac{C^2(\text{problem})}{K} \tag{4.50}$$

where $C(\text{problem})$ is a measure of problem complexity he defines, and K the number of hidden neurons. The interpretation is that for any continuous function of finite complexity, there exists a single hidden layer approximation with K neurons whose error is smaller than $\frac{C^2(\text{problem})}{K}$. Thus, by increasing the number K of neurons it is possible to reduce approximation errors in a linear fashion.

However, this very good approximation property does not prevent the over-fitting problem. Indeed, the practical problem is to choose the right weight values on the basis of the learning set, and here variance comes into play. Thus, Barron further shows that variance is upper bounded in the following fashion

$$\text{Variance} \leq \frac{Kn \log N}{N} \tag{4.51}$$

where N denotes the learning set size and n the number of inputs. In particular, variance is proportional to the product of the number of neurons and the input space dimensionality and decreases less than linearly with the sample size.

Summing the above upper bounds yields an overall approximation bound for single hidden layer perceptrons learned from data :

$$SE = \text{Bias}^2 + \text{Variance} \leq \frac{C^2(\text{problem})}{K} + \frac{Kn \log N}{N}. \tag{4.52}$$

Thus, one can define the optimal number of neurons minimizing this upper bound, yielding

$$K^* = C(\text{problem})\sqrt{\frac{N}{n \log N}}. \tag{4.53}$$

The latter formula is useful from a qualitative viewpoint, because it shows that the optimal number of neurons increases rather slowly with the number of learning states. Thus, the linear decrease in approximation error can only be achieved provided that the learning sample size increases more than quadratically.

For practical use, the formula needs the evaluation of $C(\text{problem})$ which is either not possible or computationally not feasible. Thus, it needs to be approximated.

Nevertheless, the ideas presented above have been practically exploited in the following way : with a few repetitive trials on samples of small and medium size it is possible to determine the mean optimal number of neurons for a few values of N. Then eqn. (4.53) can be used to approximate $C(\text{problem})$, and together with eqn. (4.52) to compute the sample size that would be required to reach a given level of accuracy.

4.3.7 Structure and preprocessing

Thus, from the theoretical point of view on can say that most of the practically interesting input/output mappings can be well approximated with a single hidden layer. On the other hand, more than two hidden layers are seldom considered in practice.

For a single hidden layer perceptron the total number of weights w is equal to $(n+r+1)K$, where n, r, h denote respectively the number of input attributes, output variables and hidden neurons. An often used rule of thumb consists of choosing K so as to obtain a number of weights w equal to the number of learning samples divided by a constant factor, say five to ten, so as to ensure a high enough redundancy in the learning set and reduce over-fitting.

Data preprocessing mainly consists of scaling the input attributes, so as to avoid saturating the nonlinear activation functions during the initial iterations of the back-propagation process. Such a saturation would lead to a flat SE behavior and the possible freezing of the network weights to their initial values. In the context of classification problems, we have generally used the -1/+1 output encoding, using one output neuron per class. In the context of regression problems, for example when trying to fit the output to a security margin, we have observed that the proper scaling of the output information not only improves the speed of convergence but also the quality of the solution.

Another interesting possibility consists of using the hybrid approach discussed in §6.3 to determine the appropriate input attributes and the structure of the multilayer perceptron on the basis of a decision tree previously built and converted.

Finally, various other network growing and pruning techniques have been proposed in the literature to determine the appropriate structure of a network. However, we refer the interested reader to [HKP91] and [Hay94] for a description of current research on this topic.

4.4 RADIAL BASIS FUNCTIONS

One of the main drawbacks of MLPs is their rather slow iterative learning algorithms. Radial basis functions (RBFs) can be seen as an alternative, often effective, single hidden layer structure which uses a faster direct learning algorithm. Below we give a very short and simplified description of radial basis functions. We refer the interested reader to [Hay94] for a much more in-depth treatment.

4.4.1 Description

Radial basis functions have a single hidden layer, which uses generally Gaussian activation functions, and a linear output layer. The hidden layer neurons compute the generalized distance from the input vector to some prototypes (one for each hidden neuron). The output layer then acts as computing an average value from values stored for each hidden neuron and these distances. Thus radial basis functions act in a similar way to nearest neighbor regression or Parzen methods described in Chapter 3.

A Gaussian RBF corresponds to a regression model of the following form

$$r(\boldsymbol{a}) = \sum_{i=1,K} \boldsymbol{v}_i \exp\{-\frac{||\boldsymbol{a} - \boldsymbol{a}^i||^2}{2\sigma^2}\}, \tag{4.54}$$

where the vectors $\boldsymbol{a}^i$ represent the location of the centers and $\boldsymbol{v}_i$ are the output vectors attached to each neuron.

For example, if an object is exactly located at center $\boldsymbol{a}^i$ the above sum would reduce to

$$r(\boldsymbol{a}^i) = \boldsymbol{v}_i + \sum_{j \neq i} \boldsymbol{v}_j \exp\{-\frac{||\boldsymbol{a}^i - \boldsymbol{a}^j||^2}{2\sigma^2}\} \tag{4.55}$$

and assuming that the distances between the different centers are large this would yield

$$r(\boldsymbol{a}^i) = \boldsymbol{v}_i + \epsilon. \tag{4.56}$$

In other words, the RBF can be seen as a recalling mechanism of values attached to prototypes. The closer an object to a prototype, the closer is the value recalled to the value attached to this prototype.

4.4.2 Learning algorithms

Radial basis functions are generally learned in two direct and successive steps : (i) determination of the number and location of the centers of the Gaussian activation functions of the hidden neurons; (ii) determination of the output values attached to each hidden neuron.

Location of the hidden neurons. There are mainly two different approaches.

The first one uses a very large number of hidden neurons located randomly in the input space, without any adaptation to the problem at hand. For example a random subset of the $\boldsymbol{LS}$ could be used as centers.

The second approach consists of using an unsupervised clustering algorithm to determine a sequence of cluster centers in the input space. For example, the methods described in §3.6.2 or the Kohonen algorithm, discussed in the next section, can be used for a given number of hidden neurons. Although it is a little less direct, this approach allows one to reduce the number of hidden neurons significantly in practice.

Once the centers $\boldsymbol{a}_i$ of the Gaussian activation functions have been determined, it remains to fix their radius σ; in the most simple case it is fixed to $\frac{d}{\sqrt{2k}}$ where d denotes the maximum distance between the chosen centers and K is the number of hidden neurons (i.e. of centers).

Learning the output vectors. Once centers and radius have been determined, the hidden layer can be viewed as providing a nonlinear transformation of the input space into a new feature space of dimension K.

The output layer (materialized by the vectors $\boldsymbol{v}_i$) then consists of approximating the regression function $\boldsymbol{y}$ by a linear model in this new feature space. Thus the identification of appropriate vectors $\boldsymbol{v}_i$ on the basis of a $\boldsymbol{LS}$, may be solved by the linear least-squares method described in section §3.2.2.

Determining the number of centers. Notice that since both clustering and linear regression are rather efficient techniques, the optimal number of centers K^* can be determined by generating a sequence of RBFs for growing values of K and using cross-validation to identify the best one.

4.5 KOHONEN FEATURE MAPS

Let us turn for a brief while back to the realm of unsupervised learning, and consider one of the neural network based approaches to this problem, namely the feature maps developed by Kohonen [Koh90]. We refer the interested reader to [Pao89, Zur90, HKP91] which discuss other unsupervised ANN based methods.

There are three main reasons why we have chosen to describe the Kohonen feature mapping approach. First, it is a promising method for data analysis, due to its graphical interpretation possibilities; it could be particularly useful in the context of power system security assessment, where in-depth data analysis is of paramount importance. Second, this method is essentially *complementary* to the classical statistical techniques of unsupervised learning, presented earlier. Finally, some interesting applications of the Kohonen's feature map to power system security problems have been proposed in the literature [NG91, MT91, WJ95].

4.5.1 Unsupervised learning

The self organizing feature map (SOM) developed by Kohonen belongs to the category of competitive learning models which aim at representing a data set by a smaller number of representative prototypes. There are many possible practical motivations for this kind of approach. For example in the context of information communication, this may provide an efficient way of encoding information. In the context of data analysis, it may provide a small representative subset of states.

In comparison to the other similar methods, e.g. the clustering algorithms discussed in §3.6, the main originality of the SOM is that it allows us to organize the learned prototypes in a geometric fashion, for example on a one- or a two-dimensional regular grid or map.

Below, we will particularize our presentation to the two-dimensional case, which is the most usual one, for graphical interpretation reasons. The interested reader may refer to the paper by Kohonen [Koh90, Koh95] for a general description and an in depth presentation of the biological motivations underlying the two-dimensional SOM.

To fix ideas we have represented in Fig. 4.5 a hypothetical two-dimensional 4×6 feature map. Each neuron i, j corresponds to a prototype in the attribute space, say $\boldsymbol{a}^{i,j}$. The connection weights from the input layer to the map correspond to the attribute values of the corresponding prototype. Further, in addition to an a priori defined distance $\delta(\boldsymbol{a}^{i,j}, \boldsymbol{a}^{k,l})$ in the attribute space, the relative location of these prototypes on the feature map defines a *topological* distance.

In this model, the output corresponding to an object o is defined as the nearest prototype in the attribute space, i.e. $\boldsymbol{a}^{i^*,j^*}$ such that

$$\delta(\boldsymbol{a}(o), \boldsymbol{a}^{i^*,j^*}) \leq \delta(\boldsymbol{a}(o), \boldsymbol{a}^{i,j}), \ \ \forall\ i, j. \tag{4.57}$$

What is expected from the learning algorithm is to define the prototype vectors so as to minimize the quantization error, e.g. in the square error sense (as in the statistical clustering algorithms of §3.6), and in addition, to define the positions of these prototypes on the feature map, so as to preserve the topological properties of the original attribute space. More precisely, we expect that prototypes which are close in the original attribute space will be located close on the map.

Notice that this kind of objective is not very different from *multi-dimensional scaling*, which aims at finding a configuration of points (e.g. the prototypes) in a low-

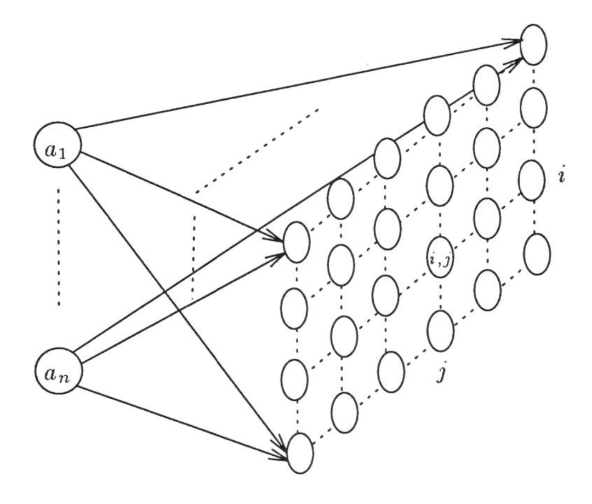

Figure 4.5. Two-dimensional Kohonen feature map

Table 4.3. Kohonen self-organizing map learning algorithm

1. *Consider the objects of the learning set in a cyclic or random sequence.*
2. *Let o be the current object,* $\mathbf{a}(\mathrm{o})$ *its attribute vector, and* $\mathbf{a}^{\mathrm{i}^*,\mathrm{j}^*}$ *its closest current prototype.*
3. *Adjust the prototype attribute vectors according to the following correction rule*

$$\left(\mathbf{a}^{\mathrm{i,j}}\right)^{\mathrm{new}} = \left(a^{i,j}\right)^{old} + \eta\Lambda\left(i-i^*, j-j^*\right)\left(a(o) - \left(a^{i,j}\right)^{old}\right). \quad (4.58)$$

dimensional space such that the distance among the points in this low-dimensional space corresponds to the distance among prototypes in the original attribute space [DH73].

Kohonen's algorithm. The elementary learning algorithm is an iterative method considering the learning set objects successively and updating the weight vectors at each step, so as to reinforce the proximity of the object and its currently closest prototypes. This is indicated in Table 4.3 in the particular case of a two-dimensional feature map.

The parameter η denotes the learning rate of the algorithm, and the function $\Lambda(\cdot,\cdot)$ is a neighborhood function, i.e. a decreasing function when the distance on the feature map increases. A frequent choice is to use the Gaussian kernel

$$\Lambda(x,y) = \exp\left\{\frac{-(x^2+y^2)}{2\sigma^2}\right\}. \tag{4.59}$$

Both the learning rate η and the width parameter σ are in practice gradually decreased during successive learning iterations. Thus, initially corrections are made so as to move a large part of the prototypes at each iteration considerably closer towards each learning object. At the later iterations, only the nearest neighbor prototype is moved and only a small correction is made at each step.

Unfortunately, the theoretical analysis of this learning algorithm has not yet been carried out very far, and among the many questions which may be raised only a few have been answered and only in the simple one-dimensional case.

Intuitively, we may feel that the above algorithm will tend to minimize a quadratic quantization error in the learning set. Of course, only a *local* minimum of this quantization error may be reached. Further, the meaning of this criterion depends, of course, on the scaling of the input attributes, in the practical case of a learning set of finite size.

On the other hand, in the case of a one-dimensional attribute space, it is possible to show that asymptotically the prototypes are regularly spaced on the feature map with an attribute density proportional to $p(a)^{2/3}$ where $p(a)$ denotes the probability density in the original attribute space. So, the Kohonen feature map tends to place the prototypes by under-sampling high probability regions and oversampling low probability ones [HKP91].

4.5.2 Possible uses

The SOM is often used for graphical data inspection and visualization.

For example, a typical application consists of building a two dimensional feature map and displaying graphical information on this map, showing class labels or attribute values in terms of the i, j coordinates. This can also be used for monitoring the position of objects on the map [NG91, MT91].

Illustration 1. Similarities among power system states. To fix ideas, we have represented in Fig. 4.6 a feature map which has been constructed for an academic voltage security example which was studied in the context of the Statlog project [MST94]. A random sample of 1250 *just after disturbance* (JAD) states was generated, and each state was characterized by 28 attribute values, corresponding to the power flows, voltages and reactive power reserves. The JAD state is a short term equilibrium which is reached by a power system after that the faster (electro-mechanical) transient have vanished, and before the on-load tap changers start reacting. It generally corresponds to 20 to 40 seconds after the disturbance occurrence, and is used in the context of

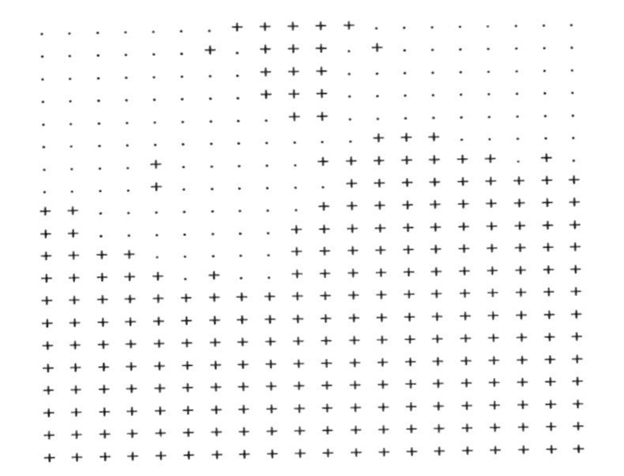

Figure 4.6. Kohonen map for the voltage security example. Adapted from [MST94]

voltage security emergency state detection. A JAD state is classified *critical* if it is expected that the subsequent action of tap-changers would drive the system towards voltage collapse, otherwise it is classified as *non-critical*.

The Kohonen map of Fig. 4.6 was determined without using information about the classification (critical vs non-critical) of the power system states. After convergence, the labels indicated in Fig. 4.6 were calibrated by determining the nearest neighbors in the learning set of each prototype vector and by associating to the latter the majority class among its nearest neighbors. In Fig. 4.6 "+" represents a prototype corresponding to a majority of critical situations, and "." a prototype corresponding to a majority of non-critical situations.

The apparent clustering in Fig. 4.6 shows, for example, that there may be two distinct types of non-critical states [MST94]. Monitoring the position on the map of the real-time power system state could provide a means to display security information to an operator. Using the latter map as a nearest neighbor classifier yields a test set error rate of 5.6%, determined in an independent test set composed of 1250 other states, generated in a similar fashion to the learning states. This is, however, a rather large error rate, since for the same problem the decision trees obtained a test set error rate of 3.8% and the multilayer perceptrons yielded error rates of 1.7%.

Illustration 2. Similarities among physical parameters. Finally, anticipating on the presentation of the voltage security study on the EDF system in §10.3, we provide an illustration of an interesting possibility of using the SOM for analyzing physical correlations among variables.

To fix ideas, let us consider the problem of defining a set of representative attributes to characterize voltage behavior of a large scale power system in the JAD state. For this problem, physical considerations suggest that the low-voltage side voltage magnitudes at the EHV/HV transformers may provide a very good picture of the severity of the disturbance and at the same time will reflect the amount of load which would be restored due to the automatic action of the transformer taps. Thus, these variables are liable to provide good indicators to detect critical situations.

However, even in a restricted region of a large scale power system, such as the one studied in §10.3, there may exist a rather large number of such transformers and correspondingly a large number of HV voltage candidate attributes.

Thus, there is a need to compress this information into a smaller number of *equivalent* voltages, or, in short there is a need to identify the *voltage coherent* regions in the power system. Once these regions are identified we may define equivalent voltages through the aggregation of elementary voltages in each region.

This is a typical attribute clustering problem, which we may try to solve with the Kohonen feature map. In our example, we start with an initial set of 39 HV voltage attributes. Each attribute is characterized by the value it assumes for a random sample of JAD states. For each attribute the same sample of states is used corresponding, in the case of our illustration, to a given disturbance and 100 randomly generated prefault operating states.

Thus, the "learning set" is composed of 39 vectors of 100 components. These vectors are pre-whitened and the Euclidean distance used by the self-organizing learning algorithm becomes equivalent to the correlation coefficient. In other words, this algorithm will try to identify regions of strongly correlated voltages. To this end, we specify a 5×6 feature map which is randomly initialized, and adapted on the basis of the above learning set. After convergence, each cell corresponds to a new vector of 100 components. The map is calibrated by identifying for each one of the 39 vectors corresponding to the 39 HV voltages its nearest neighbor on the map, i.e. the prototype to which it is most strongly correlated.

The obtained clustering is represented in the right hand part of Fig. 4.7. The non-empty cells correspond to the actual 13 prototypes determined by the algorithm. Each prototype corresponds to a set of HV voltages of which it is the nearest neighbor among all prototypes defined on the SOM. The empty prototypes are those which are the nearest neighbor of no HV voltage at all. In the left hand part of Fig. 4.7 the regions corresponding to the non-empty prototypes have been represented on the one-line diagram of the EHV system of Electricité de France (EDF).

It is interesting to notice that the location on the SOM of the prototypes corresponding to the voltage coherent regions may be compared with the adjacency of these regions on the one-line diagram. For example, regions No. 10, 11, 12, 13 which are located rather far away from the voltage weak region are also grouped together and away from the other prototypes on the feature map. On the other hand, the intermediate regions No. 6, 7, 8, 9 are also located in an intermediate position on the feature map. Finally, the regions No. 1, 2, 3, 4, 5, which are at the heart of the physical voltage security problem, are located in the left hand part of both the one-line diagram and

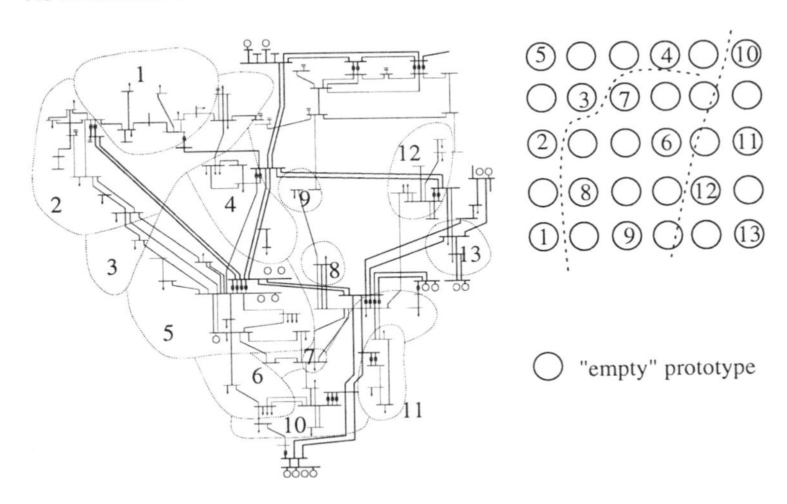

Figure 4.7. Voltage coherency SOM . Adapted from [Weh95b].

the feature map. This illustrates rather well the capability of the Kohonen SOM to preserve topological relationships among prototypes.

The above analysis, although based on a very small sample of 100 states, reflects physical information confirmed by operators' knowledge and is also consistent with other statistical analyses based on the hierarchical clustering algorithm presented in §3.6.1. With respect to this latter method, the Kohonen algorithm has several potential advantages.

First of all, in principle it is able to determine automatically the appropriate number of prototypes. In our example, this led to 13 voltage coherent regions, although the maximum possible number of prototypes in the used feature map was 30.

In addition, this method provides an indication of the topological relationship among prototypes, in terms of the distance on the feature map. For example, from the SOM in Fig. 4.7 we can make the following observations: starting at the top-left corner we first encounter region 5 (the epicenter of the disturbance), then regions 4, 7, 3, and 2 (close to the disturbance epicenter), then two groups of peripheral regions (8, 9, 1) and (6, 10, 11) and finally the remote regions 12 and 13, hardly affected by the disturbance. Thus, from top-left to the bottom-right corner of the SOM we move from strongly to weakly disturbed load regions.

We feel that this interpretability of Kohonen feature maps may be particularly useful in the context of the determination of coherent regions, where the resulting organization of prototypes may be compared with the electrical distances in the power system. The practical interest of this kind of application is further discussed in §10.3.4.

4.5.3 Supervised learning

In practice, many parameters must be tuned before obtaining good results with the above algorithm in terms of a low quantization error. This concerns first of all the choice of an appropriate map topology and neighborhood function, and a distance definition in the original attribute space. This latter is often based on an Euclidean type of distance based on some previously extracted features, e.g. a subset of pre-whitened attributes. The other choices concern parameters of the algorithm such as the success criterion and rules to define the learning rate and window, and the initial location of the prototypes. Often, several learning sessions are run in parallel on the same data set, and the most successful result is chosen as the final SOM, on the basis of the corresponding quantization error criterion.

If used correctly, the above technique may allow us to design a set of prototypes which provide a good approximation of the information contained in a learning set, as described by a set of attributes. This may directly be used for classification purposes, or similarly for regression, by calibrating the prototypes on the basis of the learning set. For example, for each prototype we may count the relative frequency of learning states of each class of which the prototype is the nearest neighbor among all prototypes on the map. These may then be used so as to associate a conditional class probability vector and a corresponding majority class.

The above is the most elementary and simplest way of exploiting a SOM for prediction. However, one may argue that this will not lead necessarily to a good behavior in terms of classification reliability, since the class information is attached a posteriori but has not been used during the construction of the map. Indeed, in practice this method turns out to provide very deceiving results in terms of classification accuracy. For example, in the StatLog study the results obtained scored worst among all methods which have been tried [MST94].

A better idea would consist of using classification information during adaptive training, so as to take into account this information to control the location of the prototypes. Applying this idea yields the so-called "learning vector quantization" (LVQ) family of methods proposed by Kohonen [Koh90], which modify the reinforcement rule of Table 4.3 so as to improve the classification reliability. We will not describe these methods in detail, but the basic idea consists of attaching a priori a class label to each prototype, and changing the sign of the $\Delta \boldsymbol{a}^{i,j}$ correction term for those prototypes which do not correspond to the same class as the current object.

4.6 FURTHER READING

In this chapter we gave some more technical details concerning the three main ANN based methods of use in the context of power systems. Our aim was to stress both similarities and differences between these methods and the statistical ones introduced in Chapter 3. Nowadays, more and more of the theoretical work on these methods aims at gaining insight into their statistical properties. However, due to space limitations we skipped most of these aspects as well as others relating the ANNs to biological brains. We refer the interested reader to the rich literature on the subject, in particular the four references given below.

Notes

1. We use the -1/+1 binary encoding rather than the often used 0/1 encoding. Further, we use also the extended attribute vector notation

$$a' \triangleq \begin{pmatrix} 1 \\ a \end{pmatrix}. \tag{4.60}$$

2. Notice that in eqn. (4.11) we take explicitly into account the dependence of states and net inputs on the object o which is presented at the MLP's input; for convenience, we will simplify this notation in most of our derivations, by leaving the dependence on o implicit.

References

[Hay94] S. Haykin, *Neural networks. A comprehensive foundation*, IEEE Press, 1994.

[HKP91] J. Hertz, A. Krogh, and R. G. Palmer, *Introduction to the theory of neural computation*, Addison Wesley, 1991.

[Koh95] T. Kohonen, *Self-organizing maps*, Springer Verlag, 1995.

[MST94] D. Michie, D.J. Spiegelhalter, and C.C. Taylor (eds.), *Machine learning, neural and statistical classification*, Ellis Horwood, 1994, Final rep. of ESPRIT project 5170 - StatLog.

[Zur90] J. M. Zurada, *Introduction to artificial neural systems*, West Publishing, 1990.

5 MACHINE LEARNING

5.1 OVERVIEW

In the broad sense, artificial intelligence (AI) is the subfield of computer science concerned with programming computers in ways that, if observed in human beings would be regarded as intelligent [Sha90] [1]. AI includes computing with symbols (crisp or fuzzy) as well as artificial neural networks. *Machine learning* (ML) is the subfield of symbolic AI concerned with learning. As most research in AI, machine learning has a twofold objective : (i) understanding learning capabilities of human beings; (ii) developing algorithms to reproduce them with the help of computers. One of the main motivations of the latter objective is the knowledge acquisition bottleneck encountered in the design of expert systems.

The aim of machine learning is to reproduce the higher level human learning process and, in particular, to build rules similar to those formulated by human experts. Hence, the main practically important characteristic of the machine learning methods discussed in this chapter is the interpretability of their results.

There are several sub-areas of machine learning, such as learning by analogy, concept formation, discovery by experimentation, explanation based learning and concept learning from examples, to mention a few ones. In this book we narrow our focus on concept learning from examples, i.e. methods which derive a symbolic description of a class of objects, on the basis of a subset of examples (and counter-examples) of this class [Mic83]. We will also consider the extension of such methods to regression problems.

Interestingly, the early work in concept learning was done by psychologists seeking to model human learning of structured concepts [HMS66]. This research has generated a whole family of *decision tree induction* systems, with the notable work on ID3 by Quinlan [Qui83] and on ACLS by Kononenko and his colleagues [KBR84, BK88]. These methods have evolved towards a set of effective and rather mature techniques, yielding commercial implementations such as the CART software [BFOS84] or the decision tree induction subroutine in the statistical package S, and the freely distributed IND package written by Buntine [Bun92]. In contrast to decision tree induction, rule learning approaches have been less successful, in particular due to their relative inefficiency in handling large scale problems [CN89, WK91, MST94].

Other machine learning researchers work on *instance based learning* (IBL) methods, which aim at developing approximate matching techniques, in order to retrieve relevant information from similar cases stored in large data bases of examples. While these methods are conceptually similar to the nearest-neighbor techniques of statistical pattern recognition, they are tailored to high level symbolic and structural example description languages [SW86, AKA91, Sal91, CS91].

Although initially most of the work in AI considered crisp binary truth values (True/False), the present trend is clearly on incorporating appropriate techniques for handling uncertainties [Che85, Pea88, BMVY93]. Within the machine learning methods, this has led to a shift from the crisp logical concept representations to the use of probabilistic or fuzzy models of attribute/class dependences. One of the important recent developments are Bayesian networks [Jen96], which allow to represent probabilistic dependences in an easily understandable graphical form.

Due to the importance of decision tree induction as a subfield of machine learning, and due to their intensive study in the context of power system security problems, we will devote most of this chapter to them.

The chapter is organized as follows : we start by introducing the general Top Down Induction of Decision Trees (TDIDT) framework (growing, pruning, and labeling of terminal nodes) which is used by almost all tree induction methods; then we describe in more detail its application to decision and regression trees, and its extension to fuzzy decision trees; finally, we end by briefly discussing some other machine learning methods and providing some hints for further reading.

5.2 GENERAL PRINCIPLES OF TREE INDUCTION

Tree induction methods have been used for nearly three decades, both in machine learning [HMS66] and in applied statistics and pattern recognition [MS63, HF69].

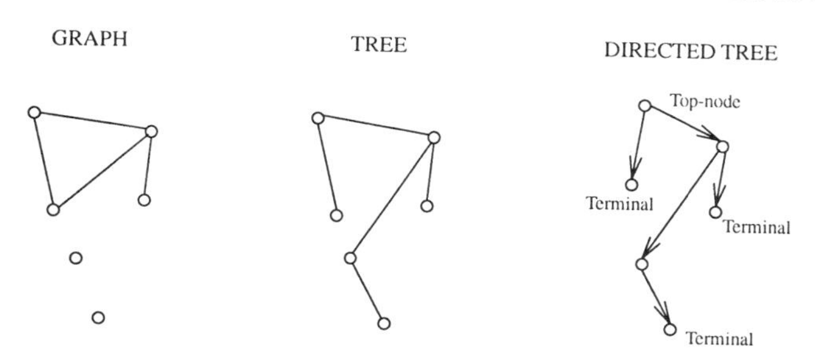

Figure 5.1. Graphs, trees and directed trees

Below we give some definitions and notation related to different types of trees, before introducing the *Top Down Induction of Decision Trees* (TDIDT) family of tree induction algorithms.

5.2.1 Trees

Graph and tree structures. A (finite) *graph* is a pair $\mathcal{G} = (\boldsymbol{\mathcal{N}}, \boldsymbol{\mathcal{A}})$ composed of a (finite) set of nodes $\boldsymbol{\mathcal{N}}$ and a (finite) set of arcs $\boldsymbol{\mathcal{A}}$, which are pairs of nodes. A graph is directed by directing its arcs.

A *tree* $\mathcal{T}$ is a connected acyclic finite graph and is directed in the following way : (i) select a node and call it the *top node* (or *root node*, denoted by $\mathcal{R}$); (ii) direct all arcs containing the top node outwards; (iii) proceed recursively, by directing arcs leaving the successor nodes of the root, until all arcs have been directed.

A non-trivial tree is a tree with at least two nodes. A node $\mathcal{N}'$ of a non-trivial tree is a successor of $\mathcal{N}$ if there is an arc $(\mathcal{N}, \mathcal{N}')$ from $\mathcal{N}$ to $\mathcal{N}'$. Except for the root node $\mathcal{R}$, every node of a directed tree is the successor of exactly one other node, called its parent node. Consequently, there is exactly one path from the root towards any other node of the tree. Graphs, trees and directed trees are illustrated in Fig. 5.1.

Nodes which have no successor nodes are called *terminal nodes*, and denoted by $\mathcal{N}_t$. Non-terminal nodes are also called *interior nodes*, and denoted by $\mathcal{N}_i$. In the sequel we will assume that, apart from terminal nodes, the set $SUCC(\mathcal{N}_i)$ of successors of an interior node contains at least two nodes.

We will denote by $DESC(\mathcal{N})$ the set of proper descendants of $\mathcal{N}$, which is recursively defined as the union of its successors and of all the descendants of these latter. The tree composed of the nodes $\{\mathcal{N}\} \cup DESC(\mathcal{N})$ and the arcs joining these nodes is called the subtree of root $\mathcal{N}$ and denoted by $\mathcal{T}(\mathcal{N})$.

Contracting a non-terminal node in a tree, consists of removing from the tree all the proper descendants of the node. A tree $\mathcal{T}'$ is said to be a pruned version of $\mathcal{T}$ if there is a subset $\boldsymbol{\mathcal{N}}'$ of non-terminal nodes of $\mathcal{T}$ such that $\mathcal{T}'$ is obtained from $\mathcal{T}$ by contracting the nodes in $\boldsymbol{\mathcal{N}}'$.

Partitioning trees. A *partitioning tree* $\mathcal{T}$ is a directed tree whose interior nodes $\mathcal{N}_i$ have been decorated with a test

$$T_{\mathcal{N}_i}(\cdot) \ : \boldsymbol{a}(\boldsymbol{U}) \longmapsto \{t_1, \ldots, t_p\}. \tag{5.1}$$

Thus, such a test is defined on the space of possible attribute values of an object, $\boldsymbol{a}(\boldsymbol{U})$. It has p possible values t_i, each one of which is associated with a unique successor, i.e. corresponds to an arc leaving the decorated node. Note that p may vary from one node to another, but is generally constrained to be small (≤ 5). In our examples will mainly consider binary trees, for which $p = 2$ at every test node. Interior nodes of a partitioning tree are also called *test nodes*.

Thus, a test allows us to direct any object from a node to one of its successors on the basis of the attribute values of the object. Consequently, starting at the top-node, any object will traverse a partitioning tree along a unique path reaching a unique terminal node.

Let us define $\boldsymbol{U}(\mathcal{N})$, the subset of $\boldsymbol{U}$ corresponding to a node $\mathcal{N}$ of $\mathcal{T}$, as the subset of objects traversing this node, while walking through the tree. Clearly, starting at the top-node and progressing towards terminal nodes, paths in the tree define a hierarchy of nested subsets :

- $\boldsymbol{U}(\mathcal{R}) = \boldsymbol{U}$, since all the paths include the top-node;

- for any interior node $\mathcal{N}_i$, the subsets of $SUCC(\mathcal{N}_i)$ form a partition of $\boldsymbol{U}(\mathcal{N}_i)$. For convenience, we will suppose in the sequel that these subsets are all non-empty;

- the subsets corresponding to the terminal nodes form a partition composed of non-empty and disjoint subsets covering $\boldsymbol{U}$.

Similarly, for any subset $\boldsymbol{X}$ of $\boldsymbol{U}$ we will denote by $\boldsymbol{X}(\mathcal{N})$ the subset $\boldsymbol{X} \cap \boldsymbol{U}(\mathcal{N})$.

Due to this correspondence, in many circumstances we will handle nodes of a partitioning tree as if they were subsets of $\boldsymbol{U}$. In particular, we will talk about the probabilities of nodes and about objects belonging to nodes.

Decision, class probability and regression trees. A *decision tree* (DT) is obtained from a partitioning tree by attaching classes to its terminal nodes. The tree is seen as a function, associating to any object the class attached to the terminal node which contains the object.

Denoting by $c(\mathcal{N}_t)$ the class associated with a terminal node, $\boldsymbol{\mathcal{N}}_{ti}$ the set of terminal nodes corresponding to class c_i, the decision regions defined by a DT are

$$\boldsymbol{D}_i = \bigcup_{\mathcal{N} \in \mathcal{N}_{ti}} \boldsymbol{U}(\mathcal{N}). \tag{5.2}$$

In the deterministic case, these subsets should ideally coincide with the classification (i.e. $\boldsymbol{D}_i = \boldsymbol{C}_i$), and the number of corresponding terminal nodes should be as small as possible for each class.

A *class probability tree* (CT) is similar to a decision tree, but its terminal nodes are decorated with conditional class probability vectors. Ideally, these (constant) probability vectors would correspond to the conditional class probabilities $p_i(\boldsymbol{U}(\mathcal{N}_t))$, in the corresponding subsets of $\boldsymbol{U}$. In addition, they should provide a maximum amount of information about classes, i.e. their residual entropy should be as close as possible to zero. This means that the tree should be designed so as to create terminal nodes where the class probabilities would ideally be independent of the attribute values of an object.

Class probability trees may easily be transformed into decision trees. For example, given a loss matrix and a probability vector, we may use the minimum risk strategy to transform probability vectors into decisions, choosing at a terminal node the class c_{j^*} minimizing the expected loss (see §2.7.7)

$$\sum_i L_{ij} p_i(\mathcal{N}_t). \tag{5.3}$$

However, in some situations it may be preferable to preserve the detailed information about conditional class probabilities, in particular when the loss matrix may change in time. Notice also that it is useful to decorate also the interior nodes with conditional class probabilities.

Finally, for *regression trees* (RT) the information stored at the nodes should describe the conditional distribution of the regression variable, typically in the form of an estimate of its expected value and (co)variance (matrix).

Notice that class probability trees may be seen as a special kind of regression trees, where the regression variable $\boldsymbol{y}$ is the class indicator variable, defined by

$$y_i(o) = \delta_{c(o),c_i}, \quad \forall\ i = 1, \ldots, m. \tag{5.4}$$

$E\{\boldsymbol{y}|\boldsymbol{X}\}$ is then equal to the conditional class probability vector $\boldsymbol{p}(\boldsymbol{X})$.

However, real regression problems are generally characterized by smooth input/output relationships, whereas class probabilities may vary in a quite discontinuous fashion, in particular in the context of deterministic problems. Further, in the case of regression problems the value of $\boldsymbol{y}$ generally varies continuously, while the class indicator variable may assume only a finite number (m) of discrete values.

In addition to the above types of trees, more sophisticated hierarchical models may be obtained by using more complicated test and terminal node decorations. For example, one may use partial propagation functions at test nodes and more elaborate models to derive information from the attribute values of objects at terminal nodes. This is illustrated below in §5.5 for fuzzy trees[2].

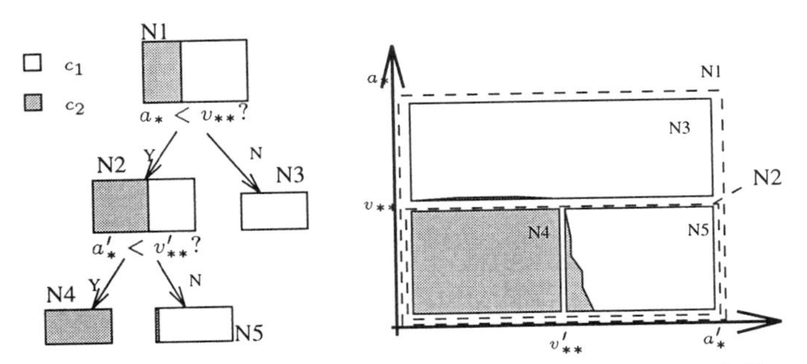

Figure 5.2. Example tree and attribute space representation

Table 5.1. Rules corresponding to the tree of Fig. 5.2

Rule N3 : *if* $[a_*(o) \geq v_{**}]$ *then* $[P(c(o) = c_1) = 1]$
Rule N4 : *if* $[a_*(o) < v_{**}]$ *and* $[a'_*(o) < v'_{**}]$ *then* $[P(c(o) = c_2) = 1]$
Rule N5 : *if* $[a_*(o) < v_{**}]$ *and* $[a'_*(o) \geq v'_{**}]$ *then* $[P(c(o) = c_1) = 1 - \epsilon]$

Figure 5.2 illustrates in its left part a simple two-class probability tree, and in its right part the corresponding sub-regions in the two-dimensional attribute space. The relative size of the white and grey parts of each node represent the conditional class probabilities estimated at this node. The relative size of a box gives an indication of the probability of belonging to the corresponding region of the attribute space. The grey shaded area in the right part of Fig. 5.2 shows the actual region in the attribute space corresponding to class c_2. Anticipating on a later discussion, we note that region N3 is not perfectly class pure although the terminal node N3 estimates $p_2 = 0$; this illustrates the possible biased character of probability estimates of trees.

Such a tree may be used to infer information about the class of an object, by directing this object towards the appropriate terminal node. Starting at the top-node (N1), the attribute test $a_* < v_{**}$? corresponding to this node is applied to the object, which is directed towards the successor node corresponding to the outcome. At each test node a particular attribute value is tested and the walk through the tree stops as soon as a terminal node is reached. This will correspond to the elementary subset in the attribute-space comprising the object and the information stored there (e.g. class probabilities, expected value of the regression variable, majority class) is extrapolated to the current object.

A tree may be translated into a complete set of non-overlapping (mutually exclusive and exhaustive) rules corresponding to its terminal nodes. For example, the translation of the tree of Fig. 5.2 is given in Table 5.1.

5.2.2 *Tree hypothesis space*

In the context of classification or regression problems, we may define a hypothesis space of trees by defining a space of candidate test functions to be applied at the interior nodes, and a class of "models" (probability, classification, regression) to be attached to the terminal nodes.

Although most of the implementations of TDIDT methods use - on purpose - simple, rather restrictive hypothesis spaces, it is important to note that these methods may be in principle generalized to more powerful hypothesis spaces.

The first restriction generally put on the test functions is that they use only a single candidate attribute at one time, the reasons for this being efficiency of search (see below) and comprehensibility of the resulting tree. Thus, the definition of test functions reduces to the definition, for each candidate attribute, of a set of candidate partitions of its possible values. This is done in a generic fashion, defining types of attributes, and, for each type, a set of candidate splits.

Symbolic, purely qualitative attributes. A purely qualitative attribute represents information which is unstructured, i.e. the values of which may not be further compared among themselves.

If $a(\boldsymbol{U}) = \{v_1, \ldots, v_q\}$ is the set of possible values of such an attribute, then, in principle, for any $p \in [2, \ldots, q]$ all possible partitions into p non-empty subsets may be used as test functions. In practice only the two extreme cases, $p = 2$ and $p = q$, have been explored in the literature.

The option $p = 2$ is often preferable, since it is found to produce simpler, and more easily interpretable trees. This leads to $(2^{q-1} - 1)$ different candidate tests of the type

$$a(o) \in \boldsymbol{V} \ ? \tag{5.5}$$

where $\boldsymbol{V}$ is a non-empty subset of $a(\boldsymbol{U})$.

Unfortunately, the exponential growth of the number of candidate splits with q makes the traditional approach, consisting of enumerating and testing each candidate partition, questionable for values of q larger than say, 10 or 20. To handle qualitative attributes with a larger number of possible values, suboptimal heuristic search must be used in the optimal splitting procedure (see the discussion by Breiman et al. [BFOS84] and Chou [Cho91]). Notice also that clustering techniques could be used in order to group the values of qualitative attributes into a given number of subsets.

Symbolic, hierarchically structured attributes. Quite often, symbolic information concerns attributes such as shape or texture, the values of which are hierarchically structured. As is illustrated in Fig. 5.3, at each node of the hierarchy a small number of subclasses of possible values are defined.

Thus, candidate partitions may be defined at a given node of a tree by identifying the most specific subset in the hierarchy containing all values assumed by the attribute

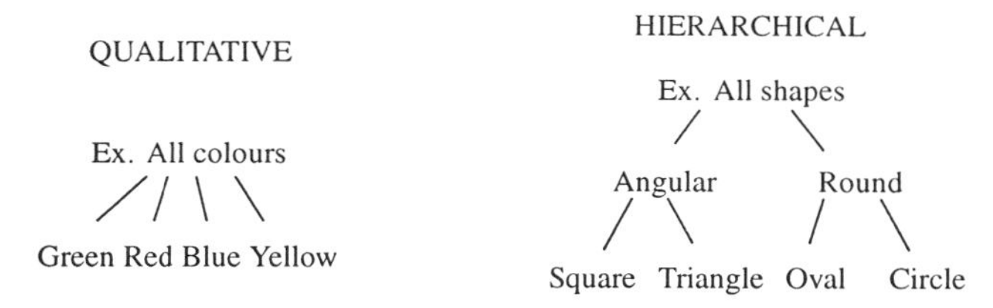

Figure 5.3. Partitioning of qualitative vs hierarchical attributes.

at this node. Only the direct subclasses of this subset will be used to define candidate partitions, which consist of adapting the "grain" of the attribute partitions to the considered subset of objects.

Let us for example consider the case of the shape attribute illustrated in Fig. 5.3. At a tree node containing all kind of objects we would use only the top-level partition distinguishing between "round" and "angular" shapes. On the other hand, at a node containing only "round" objects, we could use the distinction "circle" vs "oval" in order to split.

Ordered (integer, real valued or symbolic) attributes. Finally, a frequent kind of structure in attribute values concerns value *ordering* as it is for example the case for numerical attributes, used in most of the power system problems.

In this case, some thresholds v_i^{th} are defined yielding dichotomous tests of the form

$$a(o) < v_i^{th} \ ? \tag{5.6}$$

Some authors propose to use a small number of a priori fixed candidate thresholds [LWW+89] ; this may however lead to high quantization errors and potential loss of discriminatory information. To overcome this difficulty, a better strategy consists of adapting the candidate thresholds to the distribution of values observed in the learning set (e.g. see §5.3.3).

5.2.3 Top down induction of trees

Quite a large number of variants of tree induction algorithms have been proposed in the past, not all of which fit perfectly to the generic TDIDT procedure which we will describe below. In the next section, we will give some bibliographical information on the variants which seem most relevant to us.

The basic TDIDT procedure is a greedy algorithm, building a tree in a successive refinement approach. The implicit goal of this iterative search is to produce an as simple as possible tree, providing a maximum amount of information about the classification or the regression variable of the learning examples. For instance, the objective of the initial version of ID3 method was to build the most simple tree of minimum classification error rate in the learning set [Qui83] .

Table 5.2. Hill-climbing tree growing algorithm

Given :

- *a learning objective function : a classification* $c(\cdot)$ *or a regression variable* $y(\cdot)$*;*
- *a set of candidate attributes defined on objects* $a_i(\cdot)$*;*
- *a learning set of examples, of known attribute values and known value of the objective function;*
- *an optimal splitting rule;*
- *a stop splitting rule.*

Build : *a tree with objective function statistics at its terminal nodes : class counts of* $c(\cdot)$ *or mean and standard deviation of* $y(\cdot)$.

Procedure :

1. *create a node, attach the current learning subset to this node, and compute the objective function statistics in this learning subset;*
2. *if the stop splitting rule applies, leave this node as a terminal node;*
3. *otherwise :*
   (a) *apply the optimal splitting rule to find out the best test for splitting the current node, on the basis of the current learning subset;*
   (b) *using the above test, decompose the current learning subset into subsets, corresponding to the* p *mutually exclusive outcomes;*
   (c) *apply the same procedure to the newly created subsets.*

In the more recent approaches, the tree building decomposes generally into two subtasks : tree growing which aims at deriving the tree structure and tests, and tree pruning which aims at determining the appropriate complexity of a tree.

Tree growing. During this stage the test nodes of the tree are progressively developed, by choosing appropriate test functions, so as to provide a maximum amount of information about the output variable. The objective is to produce a simple tree of maximal *apparent* reliability.

The basic idea (see Table 5.2) is to develop one test-node after another, in an irrevocable top down fashion. The recursive tree growing algorithm starts with the complete learning set at the top-node of the tree. At each step a test function is selected in order to split the current set of examples into subsets, corresponding to the current node's successors. This process stops when no further nodes need to be developed.

This is a locally rather than globally optimal hill-climbing search, which leads to a rather efficient algorithm the computational complexity of which is at most of order

$N \log N$ in terms of the number of learning states and of order n in terms of the number of candidate attributes.

Optimal splitting. This rule defines the criterion and search procedure in order to choose the best candidate test to split the current node. Essentially the preference criterion evaluates the capacity of a candidate split to reduce the impurity of the output variable within the subset of learning states of a node. For classification problems node impurity is measured for example by Shannon entropy, while for regression problems variance is used.

Stop splitting. This rule allows us to decide whether one should further develop a node, depending on the information provided in the current learning subset. For example if the local learning set is sufficiently pure in terms of the objective function values there is no point in splitting further. Another, less obvious reason for not splitting a node is related to the so-called "over-fitting" problem, which may occur when the learning set of a node becomes too small to allow a reliable choice of a good split. This is further discussed below.

Tree pruning. The early tree induction methods merely reduce to the above growing procedure, essentially aiming at producing a maximum amount of information about the *learning* states. For example, in the context of classification it would try to split the training set into class pure subsets; in the context of regression it would try to define regions where the regression variable is constant.

Unfortunately, this simple strategy is appropriate only in the context of deterministic problems with sufficiently large learning samples, which is not the case in many practical applications. Thus it may produce overly complex, insufficiently reliable trees.

This is the so-called *over-fitting* phenomenon which may be explained intuitively. During the tree growing, the learning samples are split into subsets of decreasing size; if the method is unable to find splits which would allow us to reduce quickly the uncertainty about the objective function, these sets may become extremely small and eventually shrink to one or two examples. Indeed, provided that there exist no learning examples of different classes with identical attribute values, even a random tree growing strategy will eventually "purify" completely the learning set. Unfortunately, if the learning subsets become too small, the statistical information which may be collected from these subsets becomes unreliable. Or, stated in another way, to be able to extrapolate the statistical information collected at the terminal nodes to unseen states, these subsets must be sufficiently representative.

Thus, as for many automatic learning methods, there exists a tradeoff between the two following sources of error : *bias* which here results from insufficient splitting and *variance* which is a consequence of too much splitting. Too large trees will

Table 5.3. Hypothesis testing approach to pruning

- *Given a statistic $S(\cdot,\cdot)$ measuring the correlation of two variables.*
- *Let $f(S)$ be the sampling distribution of the statistic S under the hypothesis of statistical independence of the two variables.*
- *Given an a priori fixed risk α of not detecting the independence hypothesis, determine the corresponding threshold $S_{cr}(\alpha)$, such that*

$$\int_{S_{cr}}^{+\infty} f(S)dS = \alpha.$$

- *Estimate the value of statistic $\hat{S}^{LS}(T_N^*, y)$ applied to the objective function and the best candidate split T_N^* on the basis of the current node's learning subset.*
- *If $\hat{S}^{LS}(T_N^*, y) > S_{cr}(\alpha)$ reject the independence hypothesis, and split the node.*
- *Otherwise, accept the independence hypothesis and stop splitting.*

over-fit the data, whereas too small ones will under-exploit the information contained in the learning set. Therefore, some "smoothing" strategy is required to control the complexity of the tree and ensure that the learning samples at its terminal nodes remain sufficiently representative.

The first family of such "smoothing" strategies were actually proposed quite early in the tree induction history. Henrichon and Fu [HF69], as well as Friedman [Fri77] proposed to put a lower bound $K(N)$ on the size of the terminal nodes learning set, increasing slowly with the learning set size N, i.e. such that

$$\lim_{N\to\infty} K(N) = \infty \text{ and} \tag{5.7}$$

$$\lim_{N\to\infty} \frac{K(N)}{N} = 0. \tag{5.8}$$

The main weakness of this "naive" approach is that it takes into account only the sample size related reason for stopping development of a terminal node, and will generally lead either to overly simple or to too complex trees.

However, another reason for stopping to split a node is related to the discrimination capabilities of attributes. For example, in the extreme case where the attributes are "pure" noise, the "right" tree would be composed of a single top-node, whatever the size of the learning set. In most problems, of course, both sources of uncertainty may coexist up to a certain level, and a composite pruning criterion is required.

This consideration has yielded a second generation of pruning criteria, generally based on an hypothesis testing approach summarized in Table 5.3. Probably the first such method was proposed by Rounds, in terms of a non-parametric approach testing

Table 5.4. Tree post-pruning algorithm

1. *Define a reliability measure $R(\mathcal{T})$ (e.g. amount of information, amount of variance reduction) and a complexity measure $C(\mathcal{T})$ of trees (e.g. number of terminal nodes).*
2. *Define the global quality measure of a tree by*

$$Q_\beta(\mathcal{T}, LS) \triangleq R(\mathcal{T}, LS) - \beta C(\mathcal{T}), \tag{5.9}$$

which expresses a compromise between the apparent reliability $R(\mathcal{T}, LS)$ and complexity $C(\mathcal{T})$, the latter being more strongly penalized for large values of β.

3. *For β fixed, extract the optimally pruned tree $Pr^*(\mathcal{T}, \beta)$ of $\mathcal{T}$, such that $Q_\beta(Pr^*(\mathcal{T}, \beta), LS)$ is maximal. We will denote this as the β-optimal pruned tree of $\mathcal{T}$.*
4. *Provided that the quality measure is additive in terms of decompositions of a tree into subtrees, a simple recursive bottom up algorithm will do the β-optimal pruning.*
5. *Moreover, under the same conditions the β-optimal pruned trees form a nested sequence for increasing β; thus, if the initial tree contains k test nodes, the nested sequence of pruned tree will contain at most $k + 1$ different trees.*
6. *In particular, for $\beta = 0$ the pruned tree is the full tree; for $\beta \to \infty$, the pruned tree shrinks to a single node.*

the significance of the Kolmogorov-Smirnov distance between the class conditional attribute value distributions [Rou80]. Later on, several conceptually similar but more flexible techniques have been proposed using various χ-square like statistics [KBR84, Qui86a, WVRP89a].

Finally, the most recent and most general generation of pruning approaches consider the complexity or over-fitting control problem in a post-processing stage. In these methods, a tree is first grown completely and then simplified in a bottom up fashion, by removing its over-specified parts. The main reason for this new development was the difficulty with some of the above first and second generation stop splitting rules to adapt the thresholds (K, α, ...) to problem specifics [BFOS84]. However, we will see later on that there is a strong analogy between the stop-splitting and post-pruning approaches.

In the post-pruning approach a sequence of shrinking trees is derived from an initial fully grown one. One of these trees is then selected on the ground of its true reliability estimated honestly, e.g. using an independent test set. Various methods have been suggested [BFOS84, Qui87, Min89, Weh93a], corresponding to the algorithm given in Table 5.4.

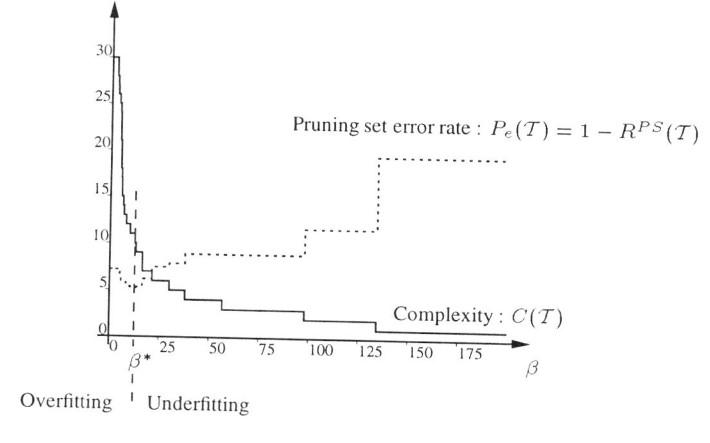

Figure 5.4. Characteristics of pruned trees for increasing β

Table 5.5. Pruned tree selection algorithm

1. *Define a tree quality measure $Q_\beta(\mathcal{T}, LS)$.*
2. *Let β increase from 0 to ∞, and generate the corresponding sequence of β-pruned trees. Since the sequence contains at most $k+1$ nested trees it may be constructed efficiently.*
3. *Compute an unbiased estimate of the latter trees' reliabilities (e.g. on the basis of an independent set of pre-classified examples PS); let $R^*(PS)$ be the corresponding maximal reliability, and let σ denote an estimate of its standard error.*
4. *Select the final tree $\mathcal{T}_s$ in the sequence as the tree of minimal complexity, and such that $R(\mathcal{T}_s, PS) \geq R^*(PS) - \sigma$, i.e. not significantly less reliable than the best one.*

Figure 5.4 illustrates a typical behavior of the complexity and the reliability (estimated on an independent set of test objects) of the optimally pruned trees, as the value of β increases from 0 to ∞. There exists an optimal value β^* of the complexity vs apparent reliability tradeoff, which leads to an optimally pruned tree of minimal estimated error rate. This overall tree selection procedure is summarized in a slightly more general version in Table 5.5, the last item of which is known as the "*1 standard error rule*".

In the sequel we will use the term *pruning set* PS, to denote the set of classified objects which is used to evaluate and select pruned trees. It is indeed necessary to distinguish this set from the true *test set* TS which is supposed to be truly *independent* of a tree, and may be used to provide unbiased estimates. Although in many practical situations the error rates obtained on the pruning set are found to be unbiased, there is no guarantee and the bias of this estimate may well depend on the pruning algorithm or on the selection rule. Clearly, the "1 standard error rule" prevents the selected tree from fitting too perfectly the PS and thus is in the favor of a low bias.

5.2.4 Extensions

Before proceeding with the description of the application of the above framework to the induction of various kinds of trees, we will briefly discuss some other interesting questions about possible variants or enhancements of the standard TDIDT method.

Variable combinations. The computational efficiency of the TDIDT algorithm is due to the fact that it searches in a reduced space of candidate trees, developing one node at a time and looking at a single attribute at a time. While this works nicely in many problems, in some situations it may be inappropriate and tend to produce very complex trees of low reliability.

In this case, one possible enhancement of the method may be to combine several attributes in a single test while splitting a node. For example, numerical attributes may be combined in a linear combination and binary attributes in a logical combination. The appropriate combinations might be chosen a priori, either manually or by using standard statistical feature extraction techniques [DH73, Fri77]. They may also be determined automatically at the time of developing a node, taking advantage of the tree growing approach to adapt the optimal combination at each tree node. Various, more or less complex strategies may be thought of in order to define an appropriate set of variables to be combined and to choose the parameters defining their combination. For example Breiman et al. propose to use a sequential forward selection procedure [BFOS84]. Utgoff [Utg88] has proposed to build decision trees with perceptrons implementing linear combinations used to predict classes at the terminal nodes; similar techniques used to define linear combinations at the test nodes are also discussed in [MKSB93].

Another, complementary possibility would be to use look ahead techniques, so as to search for high order correlations among *several* attributes and the objective function, while keeping the "single attribute per test node" representation. The latter approach would be appropriate if symbolic attributes are important.

Figure 5.5 illustrates two examples where these strategies could be useful. In its left part, a two step look ahead technique would allow us to identify the optimal decision tree, comprising four terminal nodes. The regions shown on the diagram correspond to the partition obtained by the standard (one step look ahead) TDIDT method described

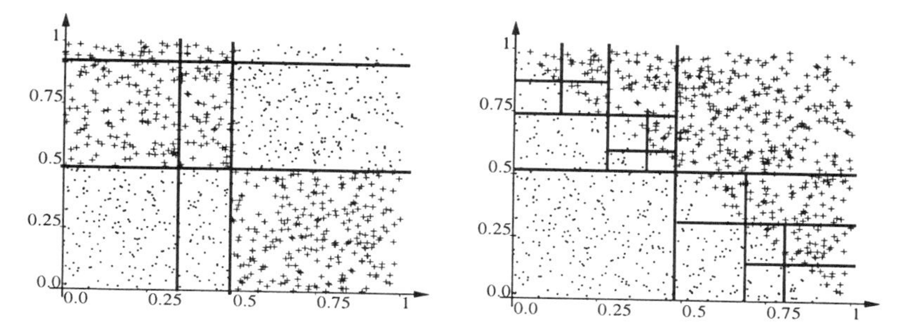

Figure 5.5. Difficult examples for the standard TDIDT approach

above. The first split at the root of the tree actually depends strongly on random variations in the learning set. Nevertheless, the resulting approximation, although overly complex, remains perfectly accurate. For example, in a simulation the tree obtained from the 1000 learning states shown on the diagram yields indeed a 100% correct classification on an independently generated test set.

For the problem illustrated in the right part of Fig. 5.5, the difficulty of the TDIDT approach is not related to its suboptimal search procedure but rather to fundamental limitations in the representation capabilities of standard trees. As shown by the stair case approximation, the resulting standard TDIDT tree is likely to be quite inaccurate. In a simulation similar to the above one, such a tree was obtained and resulted in 97-98% reliability in the test set. On the other hand extending the basic tree procedure so as to search for linear combination splits, yielding oblique trees, allowed us again to reach 100% accuracy.

Notice that several limitations exist in the above possibilities of enhancing trees. The first, rather obvious one is related to computational costs. Clearly, look ahead search time will increase exponentially with lookahead depth and is not feasible for many problems. On the other hand, the time required to determine the linear combinations rapidly increases with the number of combined attributes, but still remains feasible in practice.

Another, and at least as important drawback, is related to the fact that the interpretability of the trees will rapidly decrease if too complex tests or search criteria are used. A final limitation is due to the over-fitting problem which can be aggravated by too powerful - almost exhaustive - search techniques.

Missing attribute values. Another, often quoted practical problem occurs when attribute values are unknown for some learning states. For example, in many medical problems, attribute values determined by lengthy or potentially harmful analyses would typically be obtained only if the practitioner has good reasons to believe they will indeed

Table 5.6. Simple weighted object propagation scheme

---

1. *Let o be an object, and $w(\mathcal{N}, o)$ its weight in a tree node $\mathcal{N}$.*
2. *The weight at a successor node of $\mathcal{N}$ is obtained by $w(\mathcal{N}', o) = w(\mathcal{N}', \mathcal{N}, o)w(\mathcal{N}, o)$, where $w(\mathcal{N}', \mathcal{N}, o)$ denotes the arc strength.*
3. *Initial weights at the top-node would be usually equal to one. However, available prior information may also be used to bias initial weights.*
4. *Arc strengths are 1 for arcs corresponding to a test outcome which is known to be true for the object o, 0 for an outcome which is known to be false, and proportional to the arc probabilities otherwise.*
5. *Arc probabilities are estimated as conditional probabilities of their outcome being true, on the basis of the available examples for which the attribute value is known.*

---

provide interesting information. In these problems a high percentage of attribute values are generally unknown.

A number of methods have been proposed in order to adapt the TDIDT procedure for the treatment of unknown values. Actually there are two different situations where this problem arises. The first is when during the tree induction process some of the examples are incompletely specified. The other is when we actually use the tree to predict output information.

In both situations, a good strategy turns out to be the weighted propagation algorithm illustrated in Table 5.6. At a tree node, if the test attribute value is unknown, we estimate the probability of each outcome and we propagate the object down to every successor, along with its corresponding weight [Qui86b].

At the tree induction step, the various object counts used to evaluate probabilities or impurities are replaced by (non-integer) sums of weights. If the tree is used to predict some information for an unseen object, the latter weight is propagated through the tree and the information is collected and averaged over all relevant terminal nodes. The same technique may also be extended to the case of partial information on the attribute values[3].

A similar technique, though for a different purpose, was proposed in [CC87], where within a limited interval around the thresholds used to test continuous attributes, objects are propagated to both successors proportionally to the difference between their attribute value and the test threshold. This actually results in a kind of a *fuzzification* of the tree tests, and allows the obtained class probabilities to vary continuously, rather than in a stepwise fashion, when the attribute vector moves from one terminal node to another in the attribute space.

Generalized tree structures and tree averaging. A final possible enhancement of the TDIDT approach would be to allow more general structures than simple trees. The first possibility concerns the use of "trellis" structures allowing a node to have more than one parent node [ZAD92, Cho88, KL95].

But, let us rather discuss *option trees*[4] proposed by Buntine [Bun92].

This approach consists of inducing a *set* of trees instead of a single "optimal" one. A Bayesian framework is used to compute the posterior probability of each such tree, given its prior probability and the learning set. Finaly the predictions of the different trees are averaged proportionally to their posterior probability. Prior tree probabilities are defined as a function of the number of terminal nodes and their attached conditional class-probabilities.

More specifically, the approach proposed by Buntine consists of identifying a small number of dominant trees : those of nearly maximal posterior probability, i.e. all trees which seem to provide a reasonable explanation of the learning data. Further, the method computes, for an unseen object o, an average class probability over the dominant trees, using the following type of formula[5] :

$$P(c_i|o, \boldsymbol{LS}) \approx \frac{\sum_{\mathcal{T}} P(c_i|\mathcal{T}, o, \boldsymbol{LS}) P(\mathcal{T}|o, \boldsymbol{LS})}{\sum_{\mathcal{T}} P(\mathcal{T}|\boldsymbol{LS})}, \tag{5.10}$$

where $P(c_i|\mathcal{T}, o, \boldsymbol{LS})$ denotes the class probability predicted according to tree $\mathcal{T}$, and $P(\mathcal{T}|\boldsymbol{LS}) = P(\mathcal{T}|o, \boldsymbol{LS})$ its posterior probability, given the $\boldsymbol{LS}$ information and the current object (it is supposed to be independent of o).

Option trees are a compact representation of a set of trees which share common parts. Thus, the technique proposed in [Bun92] consists of considering several "dominant" possibilities to develop a tree node, rather than investigating only the most promising one. This includes in particular the trivial subtree, which consists of pruning the current subtree.

Thus, at each interior node of such an option tree several splits and corresponding subtrees are stored together with the corresponding posterior probability updates. During classification, an object is propagated down to each option subtree and the class probabilities inferred by the latter are averaged via a simple recursive scheme.

A similar Bayesian approach may be applied to derive smooth thresholds, and hence reduce the variance of the predictions made by decision trees [Weh97a].

5.3 CRISP DECISION TREES

In this section we illustrate the practical application of the TDIDT framework to the induction of crisp decision trees. We will describe in more detail the method developed at the University of Liège [WVRP86, WVRP89a, WP91, Weh93a] since all examples in this book have been obtained with this method.

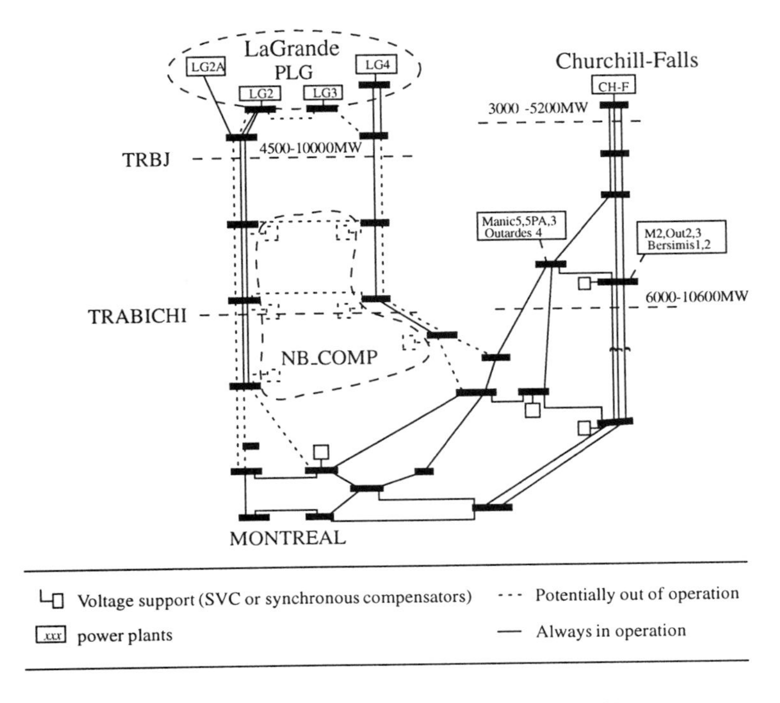

Figure 5.6. One-line diagram of 735kV Hydro-Québec system

To fix ideas, we will use throughout many numerical and graphical illustrations taken from an illustrative real life transient stability problem, introduced below. Further, we will consider only the case where the learning objective is to define decision regions or class probabilities. The generalization to regression problems and fuzzy trees are discussed in subsequent sections.

5.3.1 *Description of a real illustrative problem*

We consider the practical problem of preventive transient stability assessment of the 735kV system of Hydro-Québec, depicted in Fig. 5.6. A normal operating condition of this power system is considered as secure from the transient stability viewpoint, if it is able to withstand any permanent single-phase to ground fault, followed by line tripping, fast re-closure and final permanent tripping. It is interesting to recognize that this system, due to the very large power flows and long transmission lines, is mainly constrained by its transient stability limits.

For the sake of simplicity, we have looked at a subproblem of the overall system stability problem, considering only faults occurring within the James' Bay transmission corridor in the left part of the one-line diagram. With respect to these faults, the stability is mainly influenced by the power flows and topology within the same corridor.

For this system, a set of transient stability limits have previously been developed, in a manual approach, where operation planning engineers have determined off-line on the basis of carefully chosen simulation scenarios, a set of approximate limit tables relating the system topology and power flows to a Stable/Unstable classification. These limit tables have been implemented on the real-time computer of Hydro-Québec, via an ad hoc data base tool called LIMSEL (for LIMit SELection), which is presently in use for operation [VL86].

A data base, composed of 12497 normal operating states was generated via random sampling; it comprises more than 300 different combinations of up to 6 line outages, and about 700 different combinations of reactive voltage support equipment in operation, and a wide variety of power flow distributions. The data base generation is further discussed in §9.6.1 and a precise description of the random sampling tool developed for this purpose is given in [Weh95a].

For each state, the corresponding classification Stable/Unstable was obtained from LIMSEL running on the backup on-line computer. This yielded 3938 stable states and 8559 unstable states, among which 393 are marginally unstable and 8166 are fairly unstable.

To describe the operating states, and in order to characterize their stability, the following types of candidate attributes were computed for each state.

Power flows. The active power flow through important lines and cut-sets in the James' Bay corridor.

Power generations. Total active power generated in the 4 LaGrande (LG) power plants and various combinations.

Voltage support. The number of SVCs or synchronous condensers in operation within the six substations in the James' Bay corridor.

Topological information. Logical variables indicating for each line whether or not it is in operation in the prefault situation.

This set, composed of 67 candidate attributes was determined with the help of an expert in charge of transient stability studies at Hydro-Québec. From previous studies it was already known that the total power flow through the corridor would be an important attribute, together with the topological information and the total number of SVCs and synchronous condensers.

The histogram of Fig. 5.7 shows the statistical distribution in the data base of the total power flow in the James' Bay corridor, and the corresponding stability distribution. The height of each vertical bar represents the number of states among the 12497, for which the power flow belongs to the interval corresponding to the basis of the bar.

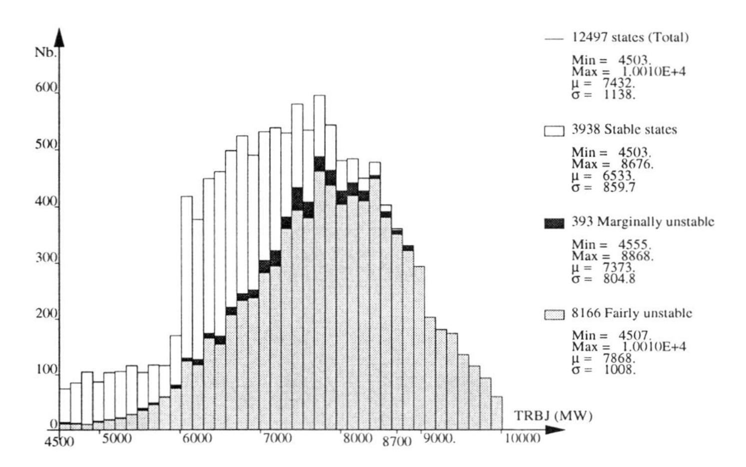

Figure 5.7. Empirical distribution of TRBJ : total James' Bay power flow

Table 5.7. Deriving classification from class probabilities

- *Let nd_i denote the non-detection cost of class c_i, i.e. the cost assigned to deciding class $c_j \neq c_i$ for an object of class c_i.*
- *Let $\hat{p}^i(\mathcal{N}_{tj})$ denote the conditional class probabilities attached to a terminal node $\mathcal{N}_{tj}$.*
- *Then, associate the decision $c_i(\mathcal{N}_{tj})$ such that the product $\hat{p}^i(\mathcal{N}_{tj})nd_i$ is maximal.*

Each bar, is further subdivided into regions of different grey shade, in proportion to the corresponding number of stable, marginal and fairly unstable states. We observe that all states which have a power flow larger than 8700 MW are unstable states, while there exist unstable states in the full range of power flows, down to 4500 MW.

5.3.2 Quality evaluation

The optimal splitting rule, as well as the stop splitting rule and the pruning criteria used in our method are derived from the entropy concept from information theory, defined in §2.7.4.

Decision trees are obtained indirectly via the construction of class probability trees and the specification of a non-detection cost vector, via the rule given in Table 5.7.

Theoretical formulation. The quality measure we have defined to evaluate class probability trees assumes the following form [Weh90, Weh92b, Weh92a, Weh93a]

$$Q(\mathcal{T}, \boldsymbol{LS}) \triangleq N I_C^{\mathcal{T}}(\boldsymbol{LS}) - \beta C(\mathcal{T}), \tag{5.11}$$

where N is the number of learning states, $I_C^{\mathcal{T}}(\boldsymbol{LS})$ apparent information (see below) and $C(\mathcal{T})$ tree complexity, which is defined as the number of terminal nodes of the tree minus one[6], so that a tree composed of a single node has zero complexity.

This quality measure, which we will "justify" heuristically below, may be derived from a theoretical *maximum a posteriori probability* (MAP) computation or equivalently from a *minimum description length* (MDL) computation, assuming either that a priori tree probabilities decrease exponentially with complexity or (equivalently) that its encoding length will require a number of bits increasing linearly with complexity. This and other more theoretical considerations are discussed in detail in the references [Weh90, Weh93a, Weh94].

A very interesting property of the quality measure is its additivity : for any decomposition of a tree into subtrees, the quality of the complete tree is equal to the sum of the qualities of the subtrees into which it is decomposed. It is a consequence of the additivity of the total information quantity and of the complexity measure.

Intuitive justification. Intuitively, the objective of building a class probability tree is to provide a maximum amount of information about the classes. This is measured by the reduction in classification entropy provided by a tree.

Let us consider a classification problem and a class probability tree $\mathcal{T}$. $H_C(\boldsymbol{X})$ denotes the initial or prior classification entropy of any subset of $\boldsymbol{U}$, defined in eqn. (2.24), and let $\{\mathcal{N}_{t1}, \ldots, \mathcal{N}_{tq}\}$ denote the partition induced on $\boldsymbol{U}$ by $\mathcal{T}$, assuming that there are q terminal nodes in $\mathcal{T}$.

Let us define the residual entropy of a tree, in a subset $\boldsymbol{X}$ of $\boldsymbol{U}$, as the expected classification entropy at its leaves

$$H_{C|\mathcal{T}}(\boldsymbol{X}) \triangleq \sum_{i=1,\ldots,q} P(\mathcal{N}_{ti}|\boldsymbol{X}) H_C(\mathcal{N}_{ti} \cap \boldsymbol{X}). \tag{5.12}$$

Then the information quantity provided by the tree in $\boldsymbol{X}$ is defined as the mutual information of the partition induced by the tree and the goal classification

$$I_C^{\mathcal{T}}(\boldsymbol{X}) \triangleq H_C(\boldsymbol{X}) - H_{C|\mathcal{T}}(\boldsymbol{X}). \tag{5.13}$$

In particular the *overall* information of a tree is obtained by replacing $\boldsymbol{X}$ with $\boldsymbol{U}$

$$I_C^{\mathcal{T}}(\boldsymbol{U}) \triangleq H_C(\boldsymbol{U}) - H_{C|\mathcal{T}}(\boldsymbol{U}), \tag{5.14}$$

and simply denoted by $I_C^{\mathcal{T}}$.

Ideally, the tree would correctly classify all the objects in $\boldsymbol{U}$ and its information would be equal to the prior entropy. In practice this is not always possible. In particular for many problems, characterized by residual uncertainty, the upper bound of information which can be provided by a tree is significantly lower than H_C.

Given a learning set, we will estimate the *apparent* information of a tree, by replacing probabilities by relative frequencies estimated in the learning set

$$I_C^{\mathcal{T}}(\boldsymbol{LS}) \triangleq H_C(\boldsymbol{LS}) - H_{C|\mathcal{T}}(\boldsymbol{LS}), \tag{5.15}$$

and the *total* apparent information quantity is obtained by multiplying the latter by the size N of the learning set.

The apparent information of a tree tends to overestimate systematically its actual information. In particular, in many circumstances it is possible to build trees with an apparent information equal to the apparent entropy, even if there is some residual uncertainty.

Intuitively, large complex trees tend to over-fit the data more strongly and their apparent information thus tends be more optimistically biased than that of smaller trees. Thus, in a quality measure it would be appropriate to compensate for this effect by penalizing in some way proportional to the tree complexity.

On the other hand, for a given tree complexity, it seems reasonable to assume that the bias of apparent information will decrease with the size of the learning set.

Hence, the quality should increase in proportion to the total amount of apparent information and decrease in proportion to the tree complexity. This suggests that the quality measure

$$Q(\mathcal{T}, \boldsymbol{LS}) \triangleq N I_C^{\mathcal{T}}(\boldsymbol{LS}) - \beta C(\mathcal{T})$$

defined in eqn. (5.11) is indeed a "reasonable" choice.

Discussion. Thus, the quality of a tree is a compromise between its complexity and its total apparent information quantity. The quality of the initial, trivial tree composed of a single root node is equal to zero, whatever the learning set, since both its complexity and its apparent information are equal to zero.

Exploiting the quality measure, for a given choice of β, the various subtasks considered in the tree induction process might be reformulated in the following way.

Growing. At each step, develop a node in such a way as to maximize the improvement of quality.

Stop splitting. Stop splitting as soon as a (local) maximum of quality is reached.

Pruning. Extract the pruned subtree of maximal quality.

In the following sections we will further discuss the variants of this approach which have been implemented.

Table 5.8. Optimal threshold identification

1. *For an attribute a and threshold v, let us denote as the left subset at a node the set of its learning states such that $a < v$ holds, and the right subset its complement.*
2. *Sort the learning subset at the current node, by increasing order of the candidate attribute considered.*
3. *Start with an empty left subset and a right subset equal to the complete learning subset of the node.*
4. *Sweep through the sorted list of states, removing at each step a state from the right subset and adding it to the left subset.*
5. *At each step, update the number of states of each class in the left and right subsets.*
6. *Let us denote by v_i the attribute value of the last object moved in the left subset; thus the left subset states are such that $a \leq v_i$.*
7. *Similarly, let us denote by v_{i+1} the attribute value of the next object to be moved, but still in the right subset; thus the right subset states are such that $a \geq v_{i+1}$ and $v_i \leq v_{i+1}$.*
8. *Only if $v_i < v_{i+1}$, we define a new candidate threshold by $v_{th} = \frac{v_i + v_{i+1}}{2}$, and compute the score of the candidate test $a(o) < v_{th}$ on the basis of the class counts in the left and right subsets.*
9. *If the score of the newly evaluated test is better than the previous optimum, we update v_{th} along with its score, as the current best test.*

5.3.3 Optimal splitting

The optimal splitting rule consists of a search for a locally optimal test maximizing a given score function. This implies finding for each candidate attribute its own optimal split and identifying the attribute which is overall optimal. This calls for the definition of a score measure and for the design of appropriate search algorithms allowing us to handle each type of candidate attributes.

We will not speak about the optimal search of binary partitions of qualitative attributes, since for power system problems the discrete attributes are generally binary topological indicators, which allow only a single partition.

However, before defining the score measure used in our algorithm, we will discuss in detail the case of numerical, essentially real valued attributes which are most important in the case of security assessment problems, as well as linear combinations of two numerical attributes which may yield an important improvement in reliability, as we will illustrate.

Optimal thresholds for ordered attributes. For a numerical attribute we proceed at each node according to the optimal threshold identification procedure described in Table 5.8 to generate the corresponding optimal partition.

Table 5.9. Linear combination search

1. *Compute the optimal threshold* v_{th1} *corresponding to* λ_1 *and* v_{th2} *to* λ_2*;* λ_1 *and* λ_2 *are specified by the user as the lower and upper bound for the search of* λ*; by default* $[-\infty \ldots \infty[$ *is used.*
2. *For each candidate value of* λ*, the corresponding threshold* $v_{th}(\lambda)$ *is determined by applying the optimal threshold search described previously to the values of the function* $a_1(o) + \lambda a_2(o)$*; the corresponding optimal score is thus determined as a function of* λ*.*
3. *The "optimal" value of* λ *is searched by using a dichotomous search in the interval* $[\lambda_1 \ldots \lambda_2[$*, with a number of iterations generally fixed a priori to less than 20.*

This search requires, about N computations of the score function in addition to the sorting the learning subset. Although it may seem bulky at first sight, it may be done rather efficiently with available computing hardware. For instance, to sort the 12497 states of our example data base of §5.3.1 with respect to the values of the TRBJ attribute, it would take about 2 seconds[7], and the overall time required to identify the optimal score within this very large subset would take about 6 additional seconds. At a tree node corresponding to "only" 1000 learning states, these times would shrink to respectively a fraction of a second and 1 second.

It is important to realize that this search procedure is applied repeatedly, for each numerical attribute and at each tree node. It will identify systematically the optimal threshold, whatever the definition of the score measure. Typically, on a 28 MIPS computer the method will not spend more, on average, than a minute at each interior node of the growing tree, even for very large learning sets and a large number of candidate attributes.

Linear combinations of attributes. In the context of power system security problems, it is frequently found that there are two important complementary attributes which share most of the information provided by a tree. In such situations, one could manually define a composite attribute as a function, or try to identify a linear combination attribute on the basis of the learning set. This amounts to identifying at a tree node a test of the form

$$a_1(o) + \lambda a_2(o) < v_{th}. \tag{5.16}$$

In our software this is done by a simple nested optimization procedure, which is indicated in Table 5.9. The computational cost of this procedure is equivalent to the treatment of about 10 to 20 real valued attributes.

An interesting generalization of the procedure would be to allow handling a higher number of attributes combined in a linear combination involving the identification of several parameters. With the above algorithm, this would, however, imply a very rapid increase in computational complexity; a more efficient numerical optimization technique should be used. In Chapter 3 we described more direct methods for devising linear models which could be combined with decision tree induction to yield oblique trees.

Remark. The above two simple search procedures may seem to be rather naive and inefficient. However, they are easy to implement and are not tied to any particular score evaluation function properties, such as continuity and differentiability. They may therefore exploit any kind of appropriate score measure.

Evaluation of candidate splits. In addition to the above described search algorithms, we need to specify the evaluation function or *score* used to select the best split. In the tree induction literature, an apparently very diverse set of measures have been proposed to select an appropriate candidate split. In [Weh96b] we discuss carefully these different measures and the purity (or uncertainty) measures from which they are generally derived. Actually, many of these apparently very different possibilities turn out to be not so different and perform rather equivalently in practice.

A convenient way to measure the impurity is to use the *entropy* function well known from thermodynamics and information theory. Among other nice properties let us mention the fact that the entropy function is the only uncertainty measure which is additive [Dar70] : the entropy of a system composed of independent subsystems is equal to the sum of the subsystems' entropies; similarly, the uncertainty of the outcome of independent events is equal to the sum of the uncertainties of each event taken alone. The other interesting thing about entropy is its probabilistic interpretation, which suggests that reducing entropy amounts to increasing posterior probabilities [Weh90, Weh92a, Weh93a] .

Thus, a simple and in practice appropriate solution consists of using the total amount of apparent information provided by a candidate partition at a node, as the criterion for selecting the most appropriate partition. This is evaluated for each test T, according to the formulas given in §2.7.4, by

$$I_C^T(\boldsymbol{LS}(\mathcal{N})) = H_C(\boldsymbol{LS}(\mathcal{N})) - H_{C|T}(\boldsymbol{LS}(\mathcal{N})). \tag{5.17}$$

Here $H_C(\boldsymbol{LS}(\mathcal{N}))$ denotes the prior classification entropy estimated in the learning subset at the node, which is obtained by

$$H_C(\boldsymbol{LS}(\mathcal{N})) = -\sum_{i=1,m} \frac{n_{i.}}{n_{..}} \log \frac{n_{i.}}{n_{..}}, \tag{5.18}$$

where $n_{i.}$ denotes the number of learning states of class c_i at the current node and $n_{..}$ its total number of learning states.

On the other hand, $H_{C|T}(\boldsymbol{LS}(\mathcal{N}))$ denotes the posterior classification entropy estimated in the learning subset at the node, given the information provided by the test. It is evaluated by

$$H_{C|T}(\boldsymbol{LS}(\mathcal{N})) = -\sum_{j=1,p} \frac{n_{.j}}{n_{..}} \sum_{i=1,m} \frac{n_{ij}}{n_{.j}} \log \frac{n_{ij}}{n_{.j}}, \tag{5.19}$$

where n_{ij} corresponds to the learning states of class c_i which correspond to the outcome t_j, and $n_{.j}$ correspond to all the states corresponding to outcome t_j.

In practice, rather than using the information quantity directly, we prefer to normalize, in order to obtain values belonging to the unit interval $[0 \ldots 1]$, independently of the prior entropy $H_C(\boldsymbol{LS}(\mathcal{N}))$. The normalized values may be interpreted as an "absolute" measure of the *correlation* between the test outcome and the classification, a value of 1 corresponding to total correlation and a value of 0 to statistical independence. In particular, information quantities obtained at different nodes of a tree, or with various classifications, may still be compared thanks to the normalization property.

In [Weh96b] we compare several possibilities to normalize the information quantity. It turns out that the resulting tree performance, in terms of complexity and reliability, is not very sensitive to the particular choice of score measure. Even in a much larger class of purity measures, not necessarily derived from the logarithmic entropy concept, the resulting tree performances remain very stable. In our method we have chosen to use the normalization I_C^T by the mean value of H_C and H_T indicated in §2.7.4, which was suggested by Kvålseth [Kvå87].

Thus our score measure is defined by

$$SCORE(T, \boldsymbol{LS}(\mathcal{N})) \triangleq C_C^T(\boldsymbol{LS}(\mathcal{N})) \triangleq \frac{2I_C^T(\boldsymbol{LS}(\mathcal{N}))}{H_C(\boldsymbol{LS}(\mathcal{N})) + H_T(\boldsymbol{LS}(\mathcal{N}))}, \tag{5.20}$$

where H_T is the uncertainty or entropy related to the outcome of the test, and is estimated by

$$H_T(\boldsymbol{LS}(\mathcal{N})) \triangleq -\sum_{j=1,p} \frac{n_{.j}}{n_{..}} \log \frac{n_{.j}}{n_{..}}. \tag{5.21}$$

Illustration. Let us consider our example problem, and let us compute the score obtained by the test $TRBJ < 7308.5MW$, used to partition the complete data base composed of the 12497 states. This test splits the data base into two subsets composed respectively of 3234 stable and 2408 unstable states for which the condition is true, and 704 stable and 6151 unstable states for which the condition is not true. This is graphically represented in Table 5.10.

Using logarithms in base two, the prior classification entropy of the complete learning set is computed by

$$H_C(\boldsymbol{LS}(\mathcal{R})) \quad = \quad -[\frac{3938}{12497}\log_2\frac{3938}{12497} + \frac{8559}{12497}\log_2\frac{8559}{12497}]$$

Table 5.10. Splitting of the data base by a test

	Stability classes		
TRBJ< 7308.5	Stable	Unstable	Total
true	n_{11}=3234	n_{21}=2408	$n_{.1}$=5642
false	n_{12}=704	n_{22}=6151	$n_{.2}$=6855
Total	$n_{1.}$=3938	$n_{2.}$=8559	$n_{..}$=12497

$H_{C|T=true} = 0.984$

$H_{C|T=false} = 0.477$

$H_C = 0.899$

$H_T = 0.993$

$$= \quad 0.899 bit,$$

and the posterior entropy is computed by

$$\begin{aligned} H_{C|T}(\boldsymbol{LS}(\mathcal{N})) &= -[\frac{5642}{12497}\left\{\frac{3234}{5642}\log_2\frac{3234}{5642}+\frac{2408}{5642}\log_2\frac{2408}{5642}\right\} \\ &\quad +\frac{6855}{12497}\left\{\frac{704}{6855}\log_2\frac{704}{6855}+\frac{6151}{6855}\log_2\frac{6151}{6855}\right\}] \\ &= 0.706 bit. \end{aligned}$$

Thus, the apparent information provided by this split is obtained by $I_C^T = 0.899 - 0.706 = 0.193 bit$.

Finally, the entropy related to the test outcome is obtained by

$$\begin{aligned} H_T(\boldsymbol{LS}(\mathcal{R})) &= -[\frac{5642}{12497}\log_2\frac{5642}{12497}+\frac{6855}{12497}\log_2\frac{6855}{12497}] \\ &= 0.993 bit, \end{aligned}$$

and thus the score associated to the above test is obtained by

$$SCORE(T, \boldsymbol{LS}(\mathcal{N})) = \frac{2 \times 0.193}{0.993 + 0.899} = 0.204 \,.$$

Parameter variance. Curves representing the score as a function of the test threshold are indicated in Fig. 5.8. The dotted line curve is obtained for a random sample composed of 1000 learning states drawn in the data base. The plain curve corresponds to the scores obtained when using the complete date base, as in the above derivation. We observe that the shape of the latter curve is much smoother than of the former.

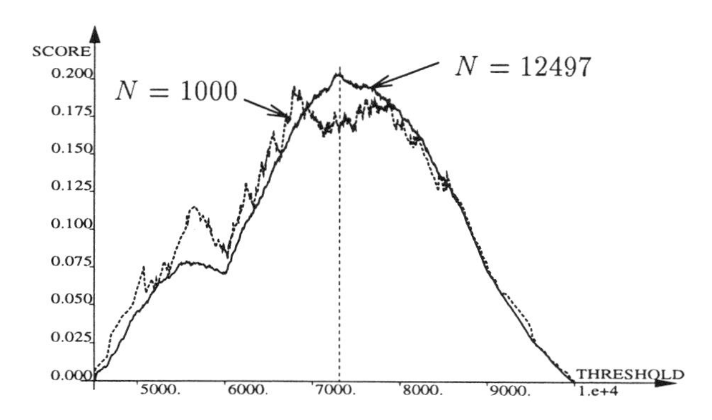

Figure 5.8. Variation of the score of the test $TRBJ < THRESHOLD$

We can also check that the value of 7308.5 MW, used in our example computation, actually corresponds to the maximum score, and thus represents the optimal threshold. On the other hand, for the dotted curve a maximum score of 0.196 is obtained for a threshold of 6767.5 MW.

The comparison of the two curves of Fig. 5.8 provides an idea of the dependence of the optimal threshold as well as the corresponding optimal score value on the random nature of a learning sample.

Therefore, it is interesting to provide information about the sampling distribution of the score measure $C_C^T(\boldsymbol{LS}(\mathcal{N}))$. This is shown to be asymptotically Gaussian and its standard deviation is estimated by [Kvå87]

$$\sigma_{C_C^T} = \sqrt{\left(\frac{C_C^T}{n_{..} I_C^T}\right)^2 \sum_{i=1,m} \sum_{j=1,p} n_{ij} \left[\log n_{ij} + \left(\frac{C_C^T}{2} - 1\right) \log(n_{i.} n_{.j}) + (1 - C_C^T) \log n_{..}\right]^2}. \quad (5.22)$$

For example, applying this formula to the test of Table 5.10 yields a standard deviation of $\sigma_{C_C^T} = 0.006$, when $N = 12497$. In the case of the optimal test obtained in the smaller random sample of $N = 1000$ of Fig. 5.8 we get a larger value $\sigma_{C_C^T} = 0.024$.

To further illustrate this random behavior, we have generated 500 random samples composed of 1000 states drawn from the above 12497. On each sample, we have computed the optimal threshold, its score and its standard deviation, according to the above theoretical formulas. The results are summarized in Fig. 5.9, which shows the empirical distributions obtained for these three parameters.

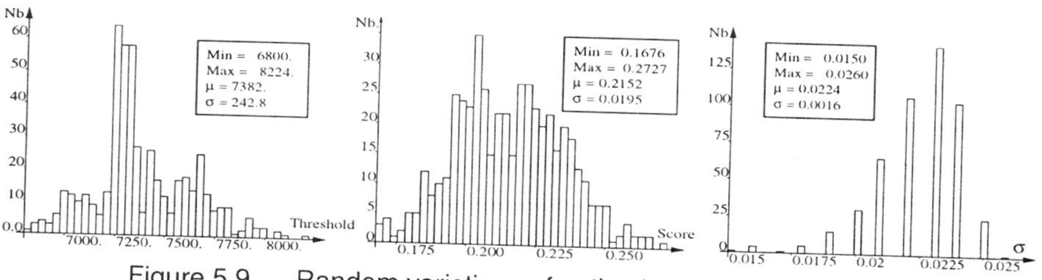

Figure 5.9. Random variations of optimal thresholds and scores

The leftmost diagram shows the distribution of the optimal threshold, which is distributed with a standard deviation of 243 MW, around the mean value of 7382 MW, close to the value of 7308.5 MW obtained in the complete data base. The central curve shows the corresponding distribution of scores and the rightmost curve the distribution of its standard deviation, computed by the above formula. We can observe that the mean value, $\mu = 0.0224$ of the latter diagram, is in good agreement with the sampling standard deviation, $\sigma = 0.0195$ observed on the central diagram. These curves illustrate the strong dependence of the parameters derived during tree building on the random nature of the learning set used. This rather high *parameter variance* is one of the main weaknesses of decision tree induction, since it can lead to difficulties in interpreting the result. Devising techniques to reduce parameter variance is an important research topic and among other possibilities fuzzy trees discussed below offer a way to reduce it (see e.g. [Weh97a]).

Comparison of candidate splits and selection. At a given test node, in order to select a test so as to develop this node, the algorithm proceeds in the following fashion.

First, for each candidate attribute it identifies the best partition for this attribute, using the appropriate search algorithm, according to the type of the attribute : discrete, ordered or linear combination of two ordered attributes.

Second, the attributes along with their optimal partitions are sorted by decreasing order of optimal score. Let $Score^*$ denote the optimal score of the best attribute, and σ^* the corresponding standard deviation computed by eqn. (5.22). The list of candidate attributes is supposed to be sorted by the user, in decreasing order of attribute preference.

Then the finally selected attribute is the first one found in the candidate attribute list, obtaining a score at least equal to $Score^* - \beta'\sigma^*$, where β' is a parameter chosen by the user. For example, using $\beta' = 0$ will always lead to selecting an attribute obtaining the highest score.

Illustration. To fix ideas, let us consider our example problem, and look at the selection of an optimal split within the complete data base and the following list of 28 candidate attributes, in the given order of preference

Power generations. PLG, PLG34, PLG2C, PLG23, PLG3, PLG4.

Global power flows. TRBJ, TRBJO, TRBJE, TRCHI, TRCHA, TRNEM, TRALB, TRABICHI, TRQMT, TRMIC, TRMAN,TRCHU.

Topological information. L7057, L7060, L7079, L7090.

Individual power flows. TR7060, TR7057, TR7079, TR7090.

Voltage support devices. NB_COMP, N_CHA.

Assuming that a value of $\beta' = 1.0$ was chosen, we obtain the information shown in Table 5.11, concerning the attribute scores in the complete data base[8]. Only the three first attributes belong to the interval of scores considered to be equivalent. Accordingly, among these the one with the highest priority in the list of candidate attributes is chosen. This is PLG, the total active power generation of the 4 LaGrande power plants (see Fig. 5.6).

Notice that the score of the two other attributes is very close. Actually, a closer look at these attributes shows that they are very strongly correlated with PLG, thus they provide similar information on the stability. TRBJ is the total power flow in the James' Bay corridor, measured nearby the generation plants and TRABICHI denotes the total power through a cross-section in the middle of the corridor. They are clearly strongly correlated with PLG.

This is confirmed by the values given in the last column of Table 5.11 which indicate the correlation of each attributes' optimal test with the optimal test of the selected attribute PLG. The correlation coefficient used here to evaluate the similarity of two tests T_1 and T_2 is defined, similarly to the the score measure, by the following formula

$$Correl(T_1, T_2) \triangleq \frac{2I_{T_1}^{T_2}(\boldsymbol{LS}(\mathcal{N}))}{H_{T_1}(\boldsymbol{LS}(\mathcal{N})) + H_{T_2}(\boldsymbol{LS}(\mathcal{N}))}. \tag{5.23}$$

Let us illustrate the use of a linear combination attribute. Although in the above table the attribute NB_COMP, denoting the total number of compensation devices in operation in the James' Bay corridor, obtains a rather low score, it is known from prior expertise that this attribute influences very strongly the stability limits of the corridor. Thus, it is presumed that a linear combination of this attribute together with the total power flow attribute TRBJ would provide increased discrimination power. Indeed, proposing this linear combination attribute to the algorithm, results in the following optimal linear combination

$$TRBJ - 227 * NB\_COMP < 5560MW \tag{5.24}$$

corresponding to a score of 0.3646, which is significantly higher than the above optimal score without linear combination attribute.

Table 5.11. Detailed information about attribute scores and correlations

```
Expanding TOP-NODE : N=12497, UNSTABLE=8559, STABLE=3938,
Total Prior Entropy N*Hc : 11234.7
  ............................
--> A test node : TOP-NODE
=============================================================
 CANDIDATE ATTR. EVALUATION:  SCORE SIGMA  N*INFO   cor PLG
=============================================================
*  TRBJ < 7307.5              0.2037 0.006  2408.8    1.00
** PLG < 7376.5               0.2037 0.006  2408.5    0.99
*  TRABICHI < 6698.5          0.2035 0.006  2401.7    0.89
-------------------------------------------------------------
   TRBJO < 4193.0             0.1437 0.006  1586.6    0.22
   TRNEM < 4257.5             0.1349 0.006  1483.6    0.22
   PLG23 < 6029.5             0.1238 0.005  1436.9    0.31
   PLG34 < 3265.5             0.0913 0.005  1082.7    0.14
   PLG4 < 1592.5              0.0727 0.004   817.6    0.11
   PLG2C < 4394.5             0.0673 0.004   787.8    0.18
   PLG3 < 1418.5              0.0653 0.004   764.4    0.09
   TRCHI < 1338.5             0.0582 0.004   475.9    0.11
   TR7060 < 956.5             0.0581 0.004   623.3    0.01
   TRCHA < 1331.5             0.0578 0.004   472.0    0.11
   TRALB < 1717.5             0.0563 0.004   495.7    0.16
   L7079 < 1.0                0.0388 0.003   346.3    0.00
   TRBJE < 2232.5             0.0376 0.003   412.9    0.10
   NB_COMP < 4.0              0.0299 0.003   277.6    0.01
   TR7057 < 1888.5            0.0235 0.002   163.8    0.05
=============================================================
CHOSEN TEST : PLG < 7376.5  (Outcomes : YES NO)
=============================================================
```

The parameters of the linear combination test, which may be rewritten as $TRBJ < 5560+227*NB\_COMP$, translate the beneficial effect of the number of compensation devices on the threshold of the total power flow attribute.

5.3.4 Stop splitting and pruning

In our initial investigations with the tree growing algorithms, in the context of transient stability assessment, we have experimented with various stop splitting criteria, using for example lower bounds on the number of learning states and/or on the residual entropy at a terminal node [WVRP86, WVRP87].

These experiments led us to the conclusion that in order to obtain good reliability it was necessary to develop the trees almost completely. But this strategy gave overly complex trees, mostly composed of random splits and which were very difficult to interpret. Thus, a strong need was felt for an approach able to distinguish among random splits and splits significantly correlated with the stability classification. We therefore proposed the hypothesis testing approach, in order to identify the situations where the apparent reduction in entropy due to a split was indeed significant [WVRP89a]. The observation that the hypothesis testing approach was equivalent to detecting a local maximum of quality became clear later, and allowed a more elegant formulation of the pruning criterion.

Stop splitting via hypothesis testing. It is important to notice that the hypothesis testing was proposed by many researchers, not the least of which is Quinlan [Qui86a], in order to handle the case of noisy attributes and noisy classifications.

Our problems however were formulated as deterministic problems, without any noise and our difficulties were related to the necessity of providing a simple approximation of a very complex problem, due to the limited amount of information provided by any learning set of reasonable size. Indeed, although we knew that a correct decision tree for most transient stability problems would be infinitely complex, we were trying to find a good compromise allowing us to represent this in a simplified fashion as far as is confirmed by the learning data.

In order to remain as coherent as possible with the strategy we used to identify the most interesting test, we decided to use the so-called $G-$statistic proposed by [Kvå87]. Indeed, one can show that under the hypothesis of statistical independence of the test issue and goal classification, the sampling distribution of the following quantity

$$G^2 \triangleq 2n_{..} \ln 2 I_C^T(\boldsymbol{LS}(\mathcal{N})), \tag{5.25}$$

which is directly proportional to the total apparent information provided by a test, follows a $\chi-$square distribution with $(m-1)(p-1)$ degrees of freedom.

Thus, conforming to the general scheme of Table 5.3, the stop-splitting rule amounts to fixing a priori a value of the non-detection risk α of the independence hypothesis and to comparing the value of $2n_{..} \ln 2 I_C^T(\boldsymbol{LS}(\mathcal{N}))$ obtained for the optimal test, with the threshold value obtained from the $\chi - square$ table. A value of $\alpha = 1.0$ would amount to systematically reject the independence hypothesis, and to consider even the smallest increase in apparent information as significant. This would yield fully grown trees, separating completely the learning states of different classes. On the other extreme, using a too small value of α would lead to develop only nodes with a very large increase in apparent information, and would produce overly simple trees.

A very large number of simulations, for a very diverse range of problems, mainly from power system transient stability and voltage security, have shown that optimal values of α are in the range of $10^{-3} \dots 10^{-4}$, which in terms of total apparent information $N I_C^T$ leads to a threshold value in the interval of $[7 \dots 15]$. These simulations have also shown that the resulting trees are generally close to optimal in terms of reliability, sometimes slightly suboptimal, but always significantly less complex than fully grown trees. To fix ideas, the ratio of the number of nodes of the full tree to the number of nodes of a pruned one with $\alpha = 10^{-4}$ lies generally between 2 and 10 [Weh90].

Thus, we conclude that the hypothesis testing approach successfully prevents trees from over-fitting their learning set, and leads to much simpler and less random trees. In terms of practical outcomes, these in general are more reliable and much easier to interpret than the trees obtained without using the hypothesis test.

Stop splitting via quality criterion. As we mentioned above, another approach to define a stop splitting criterion is based on the quality measure.

Since the objective of the hill-climbing tree growing algorithm is to maximize the tree quality, a good criterion of stop splitting would be to detect a local maximum of quality. For a given value of β, the quality variation of a tree $\mathcal{T}$ resulting from splitting a terminal node $\mathcal{N}$ with a test T, is computed by

$$\Delta Q(\mathcal{T}, \boldsymbol{LS})_{T,\mathcal{N}} \triangleq n_{..} I_C^T(\boldsymbol{LS}(\mathcal{N})) - \beta(p-1), \tag{5.26}$$

where $(p-1)$ represents the variation of the number of terminal nodes due to the node development. This is always equal to 1 in the case of binary trees. Thus, the detection of a local maximum of quality at a terminal node of a tree amounts to comparing the value of the total apparent increase in information provided by the optimal test T^*, $n_{..} I_C^{T^*}(\boldsymbol{LS}(\mathcal{N}))$, with the value of the threshold β.

Similarly to the risk α of the hypothesis testing approach, β is a user defined parameter and should be tuned according to problem specifics. A value of $\beta = 0$ would consist of not taking into account the tree complexity in the quality measure; this is equivalent to assuming $\alpha = 1.0$ in the hypothesis testing approach and produces fully grown trees. On the other hand, using very large values of β would lead to oversimplified trees. In particular, for $\beta > NH_C(\boldsymbol{LS})$ the tree will shrink to its root node, since no test will be able to provide enough information to yield an overall positive variation of Q.

Pruning and pruning sequences. There are two possible difficulties with the above described stop-splitting approaches.

The first is related to the fact that the stop-splitting criterion is only able to detect a *local* maximum of quality. As soon as a node development is not sufficiently promising, it irrevocably stops splitting this node. There are however situations were it would be possible to improve the tree, provided that at least two or more successive node developments are considered. In order words, we have reached a local maximum which is not the global maximum.

The second difficulty, which is probably more often encountered in practice, is due to the fact that the stop-splitting approaches require the user to predefine the pruning parameter, α or β, the optimal value of which may depend on problem specifics, and on the complexity vs reliability compromise which is desired.

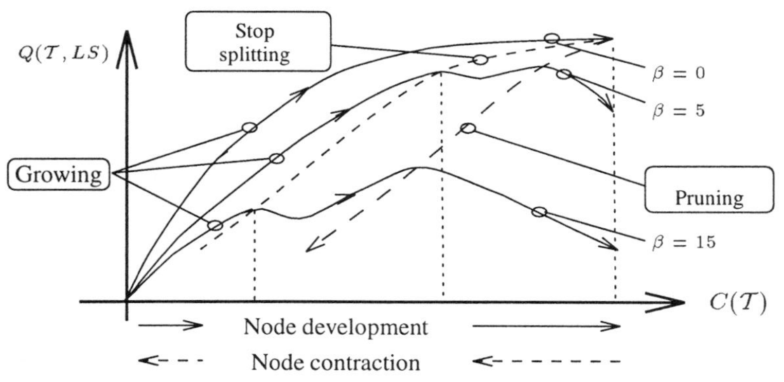

Figure 5.10. Quality variation : growing and pruning. Adapted from [Weh93a].

Each time a new learning problem is considered, the value of this parameter should be tuned to the problem characteristics. This may be done during some initial tree growing trials for various values of the parameter. Each one of the obtained trees may be evaluated on the basis of an independent test set, and the value of the parameter corresponding to the most appropriate tree would be retained for further tree building.

One of the questions of such a strategy is how often one must adapt the pruning parameter. For example, should it change as soon as the learning set size changes, or when candidate attributes are modified or only when considering a completely new learning problem. Actually, as we have already mentioned, it has been observed in practice that the optimal value of α (and thus of β) is not very sensitive to problem specifics, at least within the limited area of power system security problems, which we have studied extensively. Nevertheless, it is interesting to define a more systematic approach to identify the optimal pruning degree of a tree.

Figure 5.10 intuitively suggests the behavior of tree quality curves for variable values of β. Each one of the plain curves shows the variation of tree quality as terminal nodes are progressively developed[9]. The left hand dashed line suggests that the stop splitting approach provides a local maximum along these curves, whereas the right hand dotted line, which represents the optimally pruned trees, by definition corresponds to the global maximum along the curve. While both curves indicate that for increasing value of β the resulting tree complexity decreases, the optimally pruned tree is always of slightly higher quality and complexity than the tree obtained by the stop splitting approach.

Both of the above problems may thus be tackled by replacing the stop splitting complexity control by the tree pruning approach. This amounts to growing a tree completely, i.e. along the curve in Fig. 5.10 corresponding to $\beta = 0$ (or $\alpha = 1.0$), and then simplifying this tree by contracting its test nodes, so as to extract its pruned subtree of maximal quality, for β increasing from 0 to ∞.

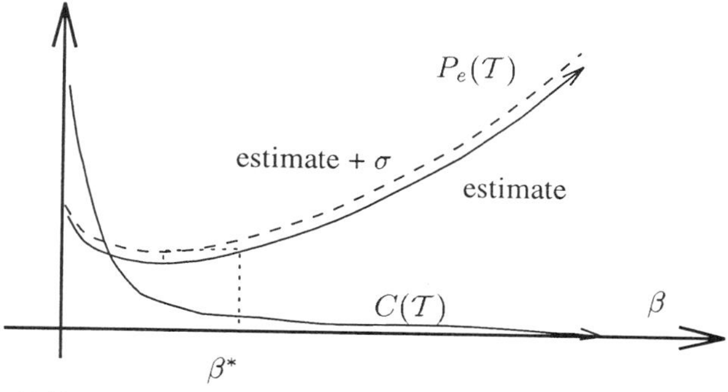

Figure 5.11. Test set error of pruned trees and "1 standard error rule"

This yields the nested sequence of shrinking trees represented by the right hand dashed line in Fig. 5.10. Using an independent pruning set to estimate the test set error rates for each one of these pruned trees will allow us to appraise their generalization capability to unseen objects. This is illustrated in Fig. 5.11, which shows the variation of the test set error rate and of the complexity of the optimally pruned trees for increasing values of β. On the basis of these latter curves, one may then select an appropriate pruning level β^*, for example by using the "1 standard error rule" as is suggested in Fig. 5.11. This consists of selecting the pruned tree as the most simple tree for which the test set error rate is not larger than the minimal test set error rate along the pruning curves plus its standard deviation.

The above algorithm may be implemented quite efficiently, allowing us to generate the complete pruning sequence with a reasonable overhead of computing time with respect to the stop splitting approach.

5.3.5 Illustration

Considering again our example transient stability problem, we have built a completely grown tree on the basis of the first 8000 states of the data base and 87 candidate attributes, including in addition to the 67 attributes proposed by the utility engineer, four linear combination attributes and some other manually defined function attributes.

This yields a very large initial tree $\mathcal{T}_0$ composed of 435 nodes, corresponding to a complexity of $C(\mathcal{T}_0) = 217^3$. This tree was evaluated on the basis of a $\boldsymbol{PS}$ composed of 2000 other states of the data base, yielding an error rate of 4.35%. Starting with this initial tree, its pruning sequence was computed and an optimal tree selected using the "1 standard error rule"[10]. The resulting tree $\mathcal{T}_{\beta^*}$ corresponds to $\beta^* \in [12.235 \ldots 12.654[$ and is composed of 115 nodes (i.e. $C(\mathcal{T}_{\beta^*}) = 57$) and has an error rate in the above test set of 3.95%.

Figure 5.12 shows the curves of the tree complexity and test set error rate along the sequence of shrinking trees for increasing values of β. For the sake of clarity the

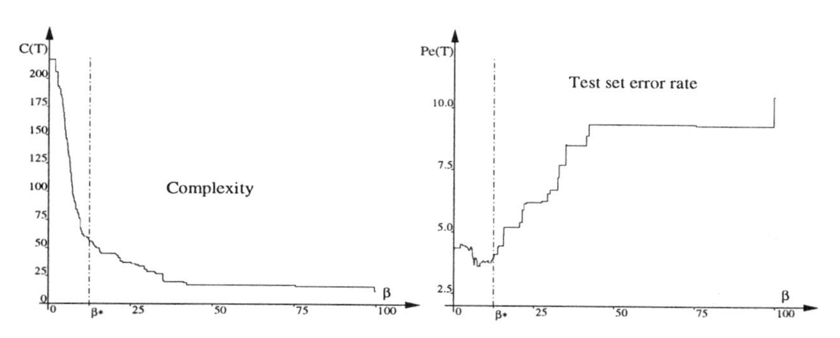

Figure 5.12. Pruning sequences for a transient stability assessment tree

graphs are zoomed on the interesting range of β values. The vertical line shows the location of the optimal tree, on the left side of which it is possible to observe the slight over-fitting of the more complex trees, which translates into an increased error rate. On the right side of this line one can observe that pruning the tree further would lead to removing some significant tests, resulting in a rapid increase in error rate.

Since the pruned tree was selected on the basis of the pruning set error rate, one could suspect the latter to be optimistically biased. Thus, we have re-tested the pruned tree, as well as the initial one, on the basis of an independent test set, composed of the 2497 remaining states of the data base. This yielded respective error rates of 4.21% for the pruned tree and 4.17% for the initial tree, which are not significantly different from the above two error rates. This is in good agreement with our overall experience, suggesting that using the "1 standard error rule" indeed produces in general quite simple trees, which are close to optimal and for which the pruning set error rate is not strongly biased. Thus in practice, it is not necessary to reserve an extra independent test set, for estimating the reliability of the pruned tree.

In terms of computational cost, the pruning approach presents an overhead with respect to the stop splitting approach, mainly due to the time required to grow the initial tree fully. In the present example, the total computing time required to grow this tree was of 3hrs 31min CPU time. Then it took about 187 seconds to generate the complete pruning sequence and to select the optimally pruned tree, and some 20 additional seconds to test the latter tree's reliability on the basis of the 2497 independent states.

In comparison, using the hypothesis testing stop splitting rule, together with a value of $\alpha = 5 * 10^{-5}$ (corresponding to the above optimal value of β) yields in this case exactly the same tree, but requires only 2hrs 16min CPU time, i.e. a reduction of about 35% with respect to the above figure.

Thus, it is of course more efficient to use the stop splitting approach, if the optimal level of pruning is known a priori. However, in order to determine this optimal value,

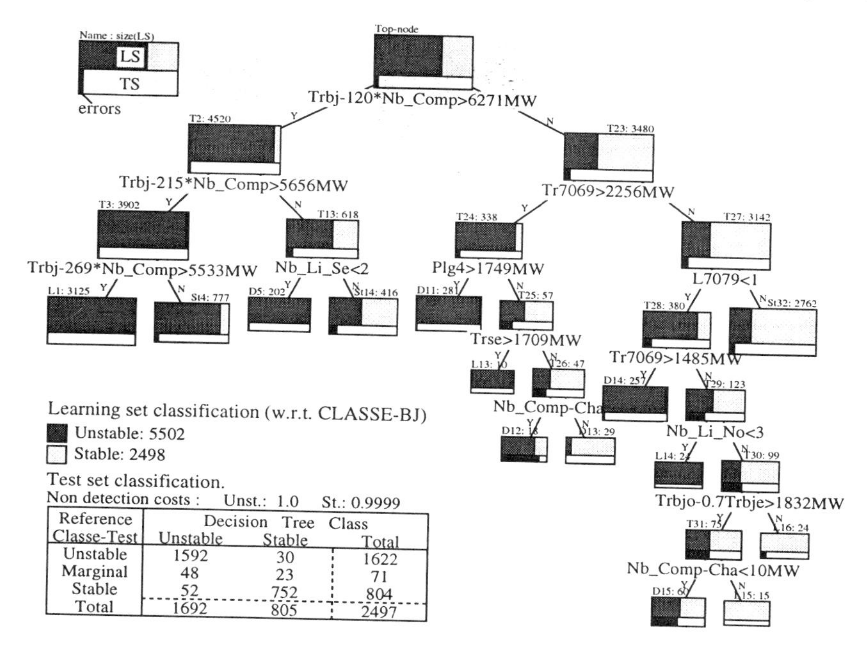

Reference Classe-Test	Decision Tree Class Unstable	Stable	Total
Unstable	1592	30	1622
Marginal	48	23	71
Stable	52	752	804
Total	1692	805	2497

Figure 5.13. Decision tree : $N = 8000$, $M = 2497$, $\alpha = 5 * 10^{-5}$, $P_e = 4.2\%$

for example during initial trials, it is much more systematic and efficient to use the pruning approach than repetitive tree building with different settings of α.

Figure 5.13 provides a partial view of the above pruned tree, showing its most important parts nearby the top-node. The notation used for a typical node is also represented at the top left hand side of the tree; one can see that each tree node is represented by a box, the upper part of which corresponds to the proportions of stable and unstable *learning* states relative to this node. In addition to the label indicating the type of a node, the number of learning states of the node is indicated next to it. Test nodes are identified by the label "Ti" or "STi", the latter corresponding to subtrees which have not been drawn on the picture. Terminal nodes are identified by a label "Li" for leafs and "Di" for deadends. A leaf is a terminal node with a sufficiently class pure learning subset, i.e. a learning subset of mean entropy lower than a predefined threshold (H_m) value taken here equal to 0.01bit, whereas a deadend is a node which corresponds to a pruned subtree.

The test results obtained when classifying the 2497 test states are shown in the table next to the tree. The non-detection costs used to assign a classification to the terminal nodes of the tree are almost identical, and the majority class is used. When there is a tie, the slightly lower non-detection cost of the stable class ensures that

Table 5.12. Percentage of NI_C^T provided by each test attribute

TRBJ+B*NB_CO :	51.8	TR7069 :	9.7	L7079 :	6.2
TRSE :	5.8	NB_COMP :	4.6	TR7062 :	4.3
NB_LI_SE :	3.4	NB_LI_NE :	1.9	PLG4 :	1.9
L7090 :	1.8	NB_COMP-CHA :	1.6	TRSO :	1.2
CLASSE-BASE :	1.0	TR7094 :	1.0	PLG+B*TRBJ :	0.8
TRNEM :	0.6	NB_LI_NO :	0.5	TR7025 :	0.4
TRNO :	0.3	TRABI :	0.3	TRBJO+B*TRBJ :	0.3
TR7044 :	0.2	PLG3 :	0.2	TR7016 :	0.2

the unstable class is systematically chosen. The table indicates the number of stable, marginally unstable and fairly unstable states, as they are classified by the tree. The 23 marginally unstable states classified stable correspond to the so-called "normal" (i.e. unavoidable) non-detections whereas the 30 fairly unstable states classified stable are the "dangerous" non-detections. The false alarms are the 52 stable states which are classified unstable by the tree. Notice that only 30 out of 1622 fairly unstable states are not detected, which yields a rather low non-detection rate of the dangerous situations of 1.85%.

In addition, at each node of the tree the proportion of erroneous classifications of test states are indicated for the corresponding subtree. At the terminal nodes this corresponds to the proportion of its test states of the minority class. At intermediate nodes, it corresponds to the mean error rate of the complete subtree, and at the top-node it corresponds to the overall error rate of the tree (i.e. 4.2%).

5.3.6 Tree and attribute information quantity

Finally, although this is hardly apparent from the above picture, we mention that the decision tree building has identified among the 87 candidate attributes 24 relevant ones. Besides, the tree allows us to reduce the initial total entropy of the learning set from $NH_C = 7166bit$ to a residual entropy value $NH_{C|T} = 965bit$. This amounts to a total information quantity provided by the tree of 86.53%.

Table 5.12 provides detailed information of the way this information is shared among the different test attributes. The attributes are sorted by decreasing values of their information quantity which is defined as the sum of the total information quantities $N(\mathcal{N})I_C^T(\mathcal{N})$ of the test nodes corresponding to a given attribute, expressed as a percentage of the total information of the tree NI_C^T.

One may observe that more than 50% of the tree information is provided by the linear combination attribute used at the top-node and at several other places in the tree. Another 40% is provided by the next eight attributes which involve the topology (L7079, NB_COMP, NB_LI_SE, NB_LI_NE) and power flows (TR7069, TRSE,

TR7062) in the James' Bay corridor as well as the active power generated in one of the LaGrande power plants (PLG4). This gives a first impression on the way various pieces of interesting information may be provided by a tree.

5.4 REGRESSION TREES

The methods for tree growing and pruning described in §5.3 may be transposed in a straightforward fashion to regression trees. The main modification consists of using a quality measure based on variance reduction instead of information quantity.

5.4.1 Variance reduction

In the context of regression, the output variable y (for the sake of simplicity of notation we will restrict our presentation to the usual case of a scalar regression variable) assumes real values and the purpose of regression tree induction is to partition the universe into regions where this variable is constant, in other words has a variance equal to zero.

Let us define, similarly to eqn. (5.12), the residual variance of a regression tree by

$$V_{y|\mathcal{T}}(\boldsymbol{X}) \triangleq \sum_{i=1,\dots,q} P(\mathcal{N}_{ti}|\boldsymbol{X}) V_y(\mathcal{N}_{ti} \cap \boldsymbol{X}) \tag{5.27}$$

where

$$V_y(\mathcal{N}_{ti} \cap \boldsymbol{X}) \triangleq E\{|y - E\{y|\mathcal{N}_{ti} \cap \boldsymbol{X})\}|^2 |\mathcal{N}_{ti} \cap \boldsymbol{X}\} \tag{5.28}$$

and $E\{y|\mathcal{N}_{ti} \cap \boldsymbol{X})\}$ denotes the expected value of y in the subset $\mathcal{N}_{ti} \cap \boldsymbol{X}$. Note that in the ideal case y would be constant at each terminal node and thus $V_{y|\mathcal{T}}$ would be equal to zero.

Then, the mean variance reduction provided by the tree in $\boldsymbol{X}$ is defined by

$$\Delta V_y^{\mathcal{T}}(\boldsymbol{X}) \triangleq V_y(\boldsymbol{X}) - V_{y|\mathcal{T}}(\boldsymbol{X}), \tag{5.29}$$

where

$$V_y(\boldsymbol{X}) \triangleq E\{|y - E\{y|\boldsymbol{X})\}|^2 |\boldsymbol{X}\} \tag{5.30}$$

is the prior variance in $\boldsymbol{X}$.

Given a learning set, we will estimate apparent variance reduction by replacing the set $\boldsymbol{X}$ by $\boldsymbol{LS}$, and define the apparent quality of a regression tree by

$$Q(\mathcal{T}, \boldsymbol{LS}) \triangleq N \Delta V_y^{\mathcal{T}}(\boldsymbol{LS}) - \beta C(\mathcal{T}), \tag{5.31}$$

as a compromise between accuracy in the learning set and complexity.

Similarly to class probability trees, the quality measure thus defined is additive and thus may be decomposed as a sum of contributions of the tree's test nodes.

5.4.2 Optimal splitting, stop splitting and node labeling

In regression trees, optimal splitting consists of searching for the test which provides the maximum amount of variance reduction in the current node's learning subset.

Stop splitting will amount to identify nodes where the variance of y is small enough (leaves) or where the amount of variance reduction is not statistically significant. The latter would call on the use of a non-parametric hypothesis test of independence. We will not elaborate on this further.

Node labeling on the other hand would consist of estimating the expected value of y in the subsets corresponding to terminal nodes. The most straightforward estimator is the mean value in the corresponding learning-subset.

The resulting method is identical to the well known regression tree algorithm given in [BFOS84].

5.4.3 Pruning

Since the quality measure defined in eqn. (5.31) is additive the pruning algorithm of Table 5.4 and 5.4 may be applied.

It should be noted that regression trees are in practice more complex than the classification counterpart. This is due to the fact that the output variable y may generally assume an infinite number of different values, yielding thus more terminal nodes containing less learning states. This is specially true if the attributes are also varying in a continuous fashion.

5.4.4 Tree and attribute information

Similarly to class probability trees, we may define the total information of the tree as the percentage of variance reduction it achieves. We may also define for each test attribute the percentage of information it provides in the tree by its contribution (summed over all nodes where it is tested) to this variance reduction.

To evaluate correlations among different optimal tests, on the other hand, we use the same information theoretic correlation measure than in class-probability trees.

5.4.5 Illustration

In power systems, regression trees have been applied successfully to the estimation of contingency severity for voltage security of the EDF system [Weh96a]. The tree in Fig. 5.14 was built for the data base described in §10.3; it tries to approximate the severity [11] of a contingency corresponding to the loss of a 400kV line in the study region.

The tree was first fully grown on the $\boldsymbol{LS}$ of size $N = 2775$ yielding a total complexity of 544 test nodes (a leave is defined to be a node where the standard

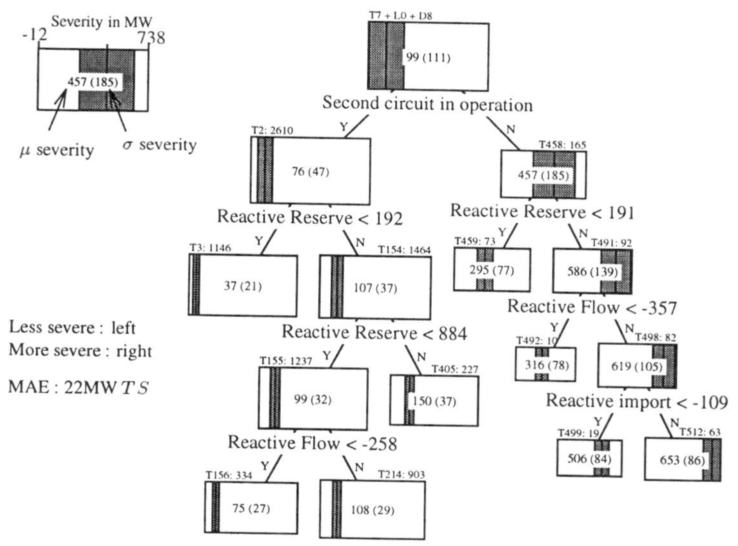

Figure 5.14. Severity regression tree : $N = 2775$, $M = 913$, $AE = 22$MW

deviation of severity is smaller than 10MW in the learning set). Then it was pruned on the basis of an independent $\boldsymbol{PS}$ (size 922) and finally tested on a further independent $\boldsymbol{TS}$ composed of $M = 913$ states.

In Fig. 5.14, the numbers inside the boxes representing the regression tree nodes indicate the mean severity in the corresponding learning subset, together with its standard deviation (in brackets). Further, the dark region represents the $\mu \pm \sigma$ interval of severities in the node's learning subset.

Thus, at the topnode the mean severity is equal to 99MW and its standard deviation is equal to 111MW. The topnode of the tree has selected a test on a topological attribute, which tests whether a line in parallel with the one tripped is in operation or not. From the information attached to the successor nodes we can appraise the variance reduction. The left successor corresponds to situations where there is a parallel line in operation, thus the contingency is less severe (in the mean 76MW) than for the right successor. The leftmost terminal node (T3) corresponds to a terminal node (actually a pruned subtree) where the mean severity is 37MW with a very small standard deviation of 21MW. For this practical example it was shown that this level of accuracy is of the same order of magnitude than that of the numerical simulation method used to compute severities.

All in all, the mean absolute error (AE) of the tree in the test set is 22MW which is fairly good, in spite of the very low complexity. Thus, this example illustrates a practical application where regression trees are well suited. Similarly to decision trees, they provide essentially interpretable information which can be analyzed physically.

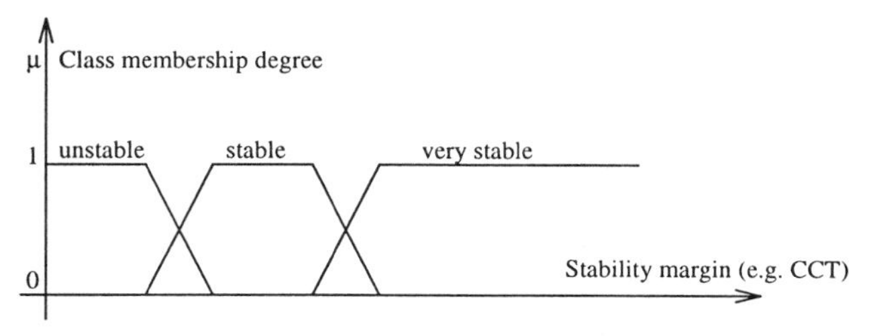

Figure 5.15. Fuzzy transient stability classes

In this context, tree pruning is paramount since it allows to reduce tree complexity and avoids representing spurious random effects.

5.5 FUZZY DECISION TREES

A very active research field concerns fuzzy decision trees, i.e. decision trees using fuzzy logic instead of classical "crisp" logic. Fuzzy logic allows one to reason about partial memberships of object to sets [DP78]. Below we will merely provide an intuitive discussion of the main differences with crisp trees.

5.5.1 Fuzzy vs crisp trees

In the context of security assessment, fuzzy classes may be defined by security margins and appropriately chosen thresholds, as illustrated in Fig. 5.15.

In order to exploit this type of information, fuzzy decision trees use fuzzy (smooth) splits at test nodes instead of crisp ones. They are therefore able to provide smooth input/output mappings, in the form of membership degrees. Figure 5.16 depicts salient differences between crisp and fuzzy trees.

A systematic approach to fuzzy tree induction is proposed in [BW95, BW96]. It is restricted to two-class problems and binary trees. Thus, it consists of using as learning target a class membership degree (e.g. a security degree derived from a security margin via **two** thresholds), and of choosing at the tree growing step at each test node an optimal attribute together with **two** thresholds, defining a transition region between left and right successors (see Fig. 5.16). In order to adapt the tree complexity automatically to the problem complexity and information at hand in the learning set, it is pruned by cross-validation.

While this technique has not yet reached maturity comparable to the crisp (decision or regression) tree approach, it appears to be quite effective in combining the data interpretation capability of symbolic machine learning with a smooth modeling capacity. In particular, the study reported in [Weh97a] shows that fuzzy trees allow to reduce parameter variance in a very effective way. Thus, fuzzy trees are at the same

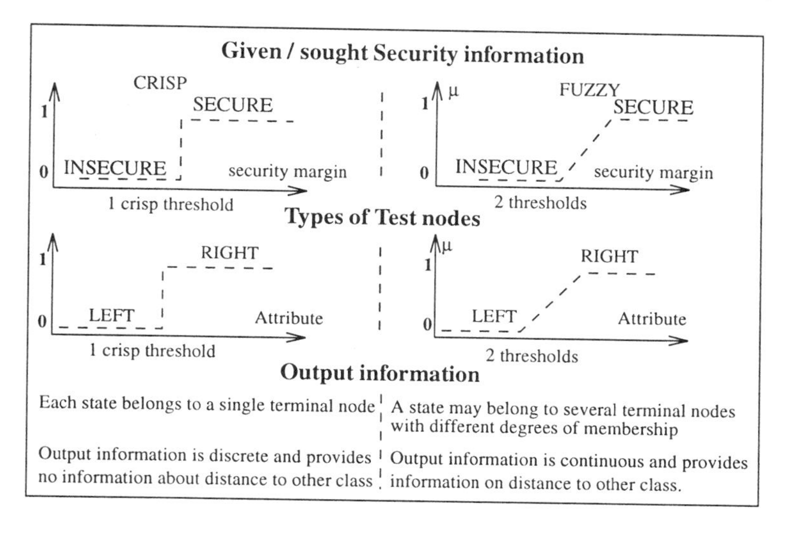

Figure 5.16. Main differences between crisp and fuzzy decision trees

time more accurate than their crisp counterpart and more stable with respect to random fluctuations of their learning set.

Nevertheless, the main drawbacks with respect to crisp trees is the much higher computational burden of the fuzzy tree learning algorithms. Further developments are necessary to make the algorithm faster so as to be of practical use in large scale problems.

5.5.2 *Illustration*

To illustrate the potentials of fuzzy decision trees in the context of security assessment, let us consider a simplified problem derived from a transient stability study carried out on the EHV system of Electricité de France (EDF) system [WPEH94].

Here the degree of stability with respect to a fault is measured by its critical clearing time (CCT). Thus, stability classes for crisp trees are defined by thresholds on the CCT.

The left hand part of Fig. 5.17 represents partially a crisp decision tree; its right hand part shows the corresponding fuzzy one. The crisp tree was built for a classification threshold of 0.240 seconds. The fuzzy one was built for a fuzzy classification, where the stability degree of a state varies *continuously* from 0 to 1 while its CCT increases from 0.215 to 0.265 seconds. While the crisp tree uses attribute thresholds to propagate a state either to right or left successors, the fuzzy tree uses transition regions defined by two thresholds, and inside the transition region a state is propagated to both

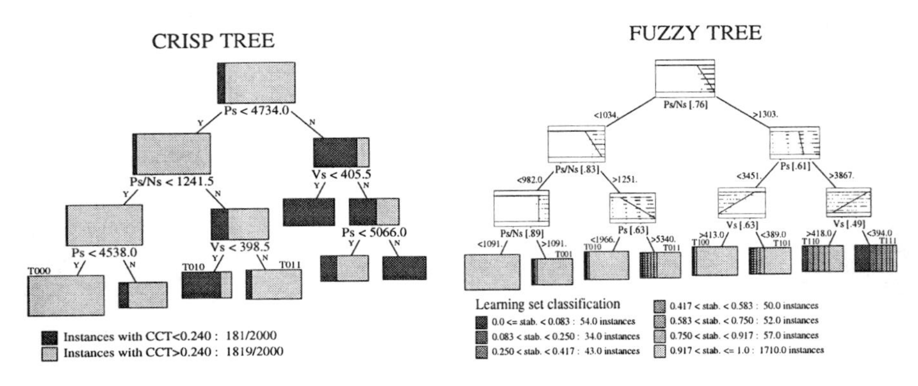

Figure 5.17. Crisp vs fuzzy decision trees

successors, with a weight varying progressively as a function of the attribute value and the thresholds.

In Fig. 5.17, each test node of the fuzzy tree represents a scatter plot of the learning states in terms of the tested attribute at this node (horizontal axis) and the degree of stability (vertical axis). Further, the transition function used to split the node is superposed. Notice that in the ideal case, this function would fit the scatter plot perfectly.

In the above example, this algorithm allowed us to improve accuracy in a significant way by reducing classification error rates from 3.3% to 1.3%, and at the same time providing more refined information about the stability of the system.

Thus, fuzzy trees are able to express continuously varying degrees of security in a very natural and effective way, similarly to smooth regression techniques. At the same time, they provide easily interpretable information, similarly to symbolic machine learning techniques. While there is still some research needed to improve the computational performances of the fuzzy tree growing and pruning algorithms, and to further validate them on different test problems, we believe that this is a very promising technique, particularly in the context of security assessment where the output information often varies continuously with input attributes.

5.6 FURTHER READING

In this chapter we have introduced the main paradigm of machine learning, namely top down induction of decision trees (TDIDT) and its variants.

While methods based on ANNs are quite similar to classical statistical techniques, these machine learning methods offer some different, unique capabilities. In particular they produce interpretable results expressed in a similar fashion to the knowledge which can be formulated by a human expert.

Because of this feature, these methods are particularly useful in the first stages of data mining applications, where one seeks to identify and understand interesting relationships found in a data base.

There are, however, several other machine learning methods which share this interpretability feature, such as rule induction, inductive logic programming, and learning in the context of Bayesian networks. Nevertheless, while all these methods show great promise, they have not yet reached maturity comparable to TDIDT. In addition, most of them suffer from difficulties in the context of real problems (large data bases, noisy information, numerical attributes, missing values ...), leading to low accuracy or prohibitive computational requirements.

TDIDT, on the other hand, scales up quite well to very large scale problems and it is also able to handle both noise and missing information. Thus, while this justifies - at the present stage of research - our focus on TDIDT, we provide below some useful references on other machine learning techniques, together with some focusing on TDIDT.

Notes

1. There are many different ways to define artificial intelligence. We recommend that the interested reader refers to [RN95], which gives a broad and modern perspective on AI.

2. In the sequel we will simply use the term *tree* ($\mathcal{T}$) to denote any kind of *decision*, *class probability*, *regression* or *fuzzy* tree.

3. The weighted object propagation scheme may be enhanced by exploiting correlations among different attributes to estimate arc probabilities. For example, Breiman et al. propose to use surrogate splits to handle missing values [BFOS84].

4. Option trees are an instance of a generic approach in automatic learning which consists of averaging several alternative models derived from a data base. This approach often allows a significant reduction of variance and hence is liable to increase the accuracy of resulting averaged predictions.

5. Note that the denominator of eqn. (5.10) would sum to 1 if all possible trees were taken into account in the sum.

6. For binary trees, the total number of nodes is related to the complexity by the formula $\#\mathcal{N} = 2C(\mathcal{T}) + 1$.

7. Within this work, illustrative CPU times are determined on a 28 MIPS SUN Sparc2 work station. Our research grade GTDIDT software was implemented in GNU Common Lisp.

8. Among the candidate attributes, L7090, L7060, TRMIC, N_CHA, TR7090, TR7079, TRMAN, TRQMT, L7057, TRCHU which obtain a score smaller than 10% of the best score are not shown in the table.

9. We suppose that at each step the node whose optimal test leads to the maximal increase in information is chosen to be developed.

10. The standard deviation of the error rate is computed by the formula $\sqrt{\frac{P_e(1-P_e)}{M}}$; in most of our simulations it is close to 0.5%

11. Severity is defined as the difference between pre-contingency and post-contingency load power margins in the study region, in MW (see §10.3.4).

References

[AKA91] D. W. Aha, D. Kibler, and M. K. Albert, *Instance-based learning algorithms*, Machine Learning **6** (1991), 37–66.

[BFOS84] L. Breiman, J. H. Friedman, R. A. Olshen, and C. J. Stone, *Classification and regression trees*, Wadsworth International (California), 1984.

[Jen96b] F. V. Jensen, *An introduction to Bayesian networks*, UCL Press, 1996.

[MD94] S. Muggleton and Luc De Raedt, *Inductive logic programming : theory and methods*, Journal of Logic Programming **19**(20) (1994), 629–679.

[Qui93] J.R. Quinlan, *C4.5. programs for machine learning*, Morgan Kaufman, 1993.

6 AUXILIARY TOOLS AND HYBRID TECHNIQUES

6.1 COMPLEMENTARY NATURE OF AUTOMATIC LEARNING METHODS

In the preceding three chapters we have described main automatic learning methods from three different paradigms : classical statistics, artificial neural nets and machine learning.

In the introduction of the book we provided motivations for a tool-box approach to automatic learning, arguing that no AL method can claim to be universal and providing illustrations of how the different methods can be complementary when facing a data mining application.

In addition to the tool-box of automatic learning methods, there are data preprocessing tools which can help to transform a problem in a more suitable representation for a particular kind of method.

Furthermore, while applying some automatic learning methods, it is often interesting to exploit information provided by some other methods, yielding various hybrid techniques which often outperform the "pure" counterparts.

Both of these aspects are discussed in this chapter.

6.2 DATA PREPROCESSING

In this section we provide some hints about a certain number of classical data preprocessing techniques, which are often used to transform an initial representation into a set of more appropriate attributes. While these methods basically belong to statistical pattern recognition, they can be useful in other kinds of applications.

They provide an intermediate tool between the manual choice of an ad hoc representation which is more of an "art", and the fully integrated automatic learning methods such as the machine learning and neural network methods.

6.2.1 Pre-whitening

Pre-whitening or normalization of data consists of linearly transforming each attribute, to obtain a zero mean and unit variance

$$a' = \frac{a - \overline{a}}{\sqrt{\overline{(a - \overline{a})^2}}}. \tag{6.1}$$

This is the least one can do in order to render numerical techniques, such as nearest neighbor computations and clustering, independent of arbitrary attribute scalings.

It is also useful in the context of multilayer perceptrons, to avoid saturating the activation functions of the first hidden layer.

6.2.2 Attribute selection

Attribute (or feature) selection consists of reducing the dimensionality of the input data by selecting a small number of the most discriminant or the most representative features. There may be two motivations for reducing the dimension of the input space. The first one is purely related to computational efficiency of the subsequent learning tasks. The second reason is more related to the problem of over-fitting.

Although there is a whole bunch of sophisticated feature selection algorithms available in the literature (see, for example [DK82]), we will only describe some basic, very simple techniques.

Attribute clustering. As we have suggested above, the clustering analysis of attributes allows us to identify groups of attributes which are very strongly correlated, i.e. which share the same physical information. In the context of power system security problems this is very frequent for variables such as power flows (see Fig. 3.7) or voltages (see §10.3). A simple dendrogram may be drawn to suggest which groups of such variables may be represented by a single prototype, e.g. a mean value. In practice, this may lead to a more efficient and more robust classification.

But since the attribute clustering technique does not take into account the classification information, it is not very selective in identifying the discriminant attributes.

Decision tree building. The next step for feature selection could be to build a decision tree, on the basis of the available pre-classified data or a regression tree, as appropriate. At each test node of the tree, the tree building algorithm provides detailed information on the score obtained by each candidate attribute and its estimated standard deviation and correlations. This makes it in general quite easy to determine a much smaller subset of the most discriminant attributes.

This technique has been illustrated in Chapter 3, in the context of the voltage security example of Table 3.1, in the discussion about the nearest neighbor rule in §3.4. This method was found to provide an important reduction in dimensionality in several other power system security applications.

Simple sequential feature selection. One method of feature selection, which has often been used for its great simplicity consists of selecting the features sequentially according to the following figure of merit

$$J \triangleq \frac{|S_B(a_1, \ldots, a_k)|}{|S_W(a_1, \ldots, a_k))|}, \tag{6.2}$$

where $|S_B(a_1, \ldots, a_k)|$ schematically represents a "between class scatter index" and $|S_W(a_1, \ldots, a_k)|$ stands for a "mean within class scatter index". Both are supposed to be computed in the attribute sub-space $(a_1, \ldots, a_k)$.

The above figure of merit can then be determined for each single attribute to choose the first attribute a_1^*, and then for each pair of attributes a_1^*, a_2, to determine the most complementary attribute a_2^*, and so on $\ldots$, yielding the sequential forward selection approach. The dual scheme consists of starting with the complete list of candidate attributes and deleting at each step the least useful one, i.e. leading to the highest value of the performance index for the remaining set of attributes.

A simplification of the above schemes consists of computing the index in a scalar attribute by attribute approach. E.g. assuming attribute independence and restriction to the two-class case, an index may be computed for each attribute in terms of the ratio

$$J_a \triangleq \frac{|\mu_{a_1} - \mu_{a_2}|^2}{p_1\sigma_{a_1}^2 + p_2\sigma_{a_2}^2}, \tag{6.3}$$

of the square difference of the class-conditional mean values to the weighted sum of the class-conditional standard deviations. Excluding strongly correlated attributes, one may select the n' best attributes according to the above criterion.

6.2.3 Feature extraction

While the feature selection methods search for an optimal *subset* of the initial attributes, the feature extraction methods aim at defining a set of - generally linear - feature combinations.

We will merely indicate the basics of the *Karhunen-Loève* expansion of *principal components analysis*. The objective of this technique is to linearly transform the initial attributes in order to concentrate the maximum amount of information in a minimum number of transformed attributes.

In the following we suppose that each attribute has been centered by subtracting its mean value. Moreover, we will use expectation operators to manipulate population quantities. The same derivations may then be applied to the finite sample case by replacing expectation operators by sample mean values.

Thus we assume that

$$E\{\boldsymbol{a}\} = \mathbf{0}. \tag{6.4}$$

Then, let us consider an orthonormal system of vectors $\boldsymbol{u}_1, \ldots, \boldsymbol{u}_n$, i.e. such that

$$\boldsymbol{u}_i^T \boldsymbol{u}_j = \delta_{ij}, \tag{6.5}$$

and express the attribute vectors as a linear combination of these vectors

$$\boldsymbol{a} \triangleq \sum_{i=1,n} \tilde{a}_i \boldsymbol{u}_i, \tag{6.6}$$

where

$$\tilde{a}_i = \boldsymbol{u}_i^T \boldsymbol{a}, \tag{6.7}$$

since the vectors $\boldsymbol{u}_i$ form an orthonormal basis.

If we take a smaller number $d < n$ of terms, truncating the above series, we obtain an approximate representation of the vector $\boldsymbol{a}$,

$$\hat{\boldsymbol{a}} = \sum_{i=1,d} \tilde{a}_i \boldsymbol{u}_i. \tag{6.8}$$

We define our "optimal representation problem" as the choice of the vectors $\boldsymbol{u}_i, i = 1, d$ which minimize the following mean squared representation error

$$\epsilon = E\left\{(\boldsymbol{a} - \hat{\boldsymbol{a}})^T (\boldsymbol{a} - \hat{\boldsymbol{a}})\right\}. \tag{6.9}$$

In other words, we search for a d-dimensional subspace of the initial attribute space, which is closest to the probability distribution in the Euclidean distance sense.

Substituting the expressions (6.6) of $\boldsymbol{a}$ and (6.8) of $\hat{\boldsymbol{a}}$ into eqn. (6.9), we obtain

$$\epsilon = E\left\{\sum_{i=d+1,n} \tilde{a}_i^2\right\}, \tag{6.10}$$

where we have exploited the orthonormality conditions (6.5).

Thus, using the expression (6.7) of the expansion coefficients, we obtain

$$\epsilon = E\left\{\sum_{i=d+1,n} \boldsymbol{u}_i^T \boldsymbol{a}\boldsymbol{a}^T \boldsymbol{u}_i\right\}, \tag{6.11}$$

or

$$\epsilon = \sum_{i=d+1,n} \boldsymbol{u}_i^T \left[E\left\{\boldsymbol{a}\boldsymbol{a}^T\right\}\right] \boldsymbol{u}_i, \tag{6.12}$$

exploiting the fact that the vectors $\boldsymbol{u}_i$ are independent of $\boldsymbol{a}$, to interchange the summation and the expectation operators. Notice that the matrix $E\left\{\boldsymbol{a}\boldsymbol{a}^T\right\}$ is the covariance matrix (cf. assumption (6.4)).

It can be shown that the stationary points of the above expression correspond to choosing for $\boldsymbol{u}_i$ eigenvectors of the covariance matrix $E\left\{\boldsymbol{a}\boldsymbol{a}^T\right\}$, i.e. such that

$$E\left\{\boldsymbol{a}\boldsymbol{a}^T\right\} \boldsymbol{u}_i = \lambda_i \boldsymbol{u}_i. \tag{6.13}$$

Under this condition the mean square representation error ϵ is computed by

$$\epsilon = \sum_{i=d+1,n} \lambda_i, \tag{6.14}$$

and will be minimal if the λ_i are chosen as the $n-d$ smallest eigenvalues of the covariance matrix.

Consequently, the optimal truncation is obtained by using the eigenvectors corresponding to the d largest eigenvalues. In principle, the truncation error would also be minimized by using any orthonormal basis of the subspace spanned by the eigenvectors corresponding to the d largest eigenvalues.

However, choosing the eigenvectors rather than an arbitrary orthonormal combination of them, has the additional feature of decorrelating the transformed attribute values. Indeed, it is easy to show that for this particular choice of $\boldsymbol{u}$ vectors $E\{\tilde{a}_i\tilde{a}_j\} = \lambda_i\delta_{ij}$.

6.3 HYBRID AUTOMATIC LEARNING METHODS

By looking at the recent literature in AL, one may observe that a large part of it is devoted to so-called hybrid or multi-strategy methods [MW97]. These approaches aim at combining different elementary AL methods in order to extract as much information from a data base as possible. While there is a large diversity of such methods, below we will describe only two, where the hybrid approach is indeed a very useful approach to enable the application of MLP and KNN based methods to real large scale problems. They have already been applied in the context of various power system problems.

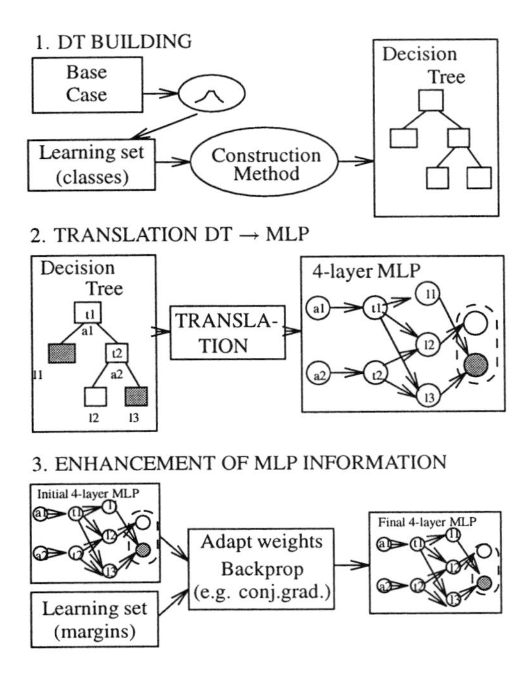

Figure 6.1. Hybrid DT-ANN approach

6.3.1 Decision trees and multilayer perceptrons

The hybrid Decision Tree - Artificial Neural Network (DT-ANN) approach aims at combining the advantages of MLPs and *DT*s while circumventing their weaknesses. DTs are used to yield a first shot, transparent and interpretable model of the relationship between attributes and output information. The powerful nonlinear mapping capacities of multilayer perceptrons are then exploited to augment the discrete classification of the tree with a continuous margin type of information. This richer information may be used in various ways; in particular, it may contribute to making better decisions during the use of the method.

Below we will describe this method in the context of power system security assessment, but it may be applied to other problems as well.

The hybrid approach is schematically shown in Fig. 6.1. Decision trees are first built using a data base composed of preclassified power system states; they identify the relevant test attributes for the security problem of concern, and express, in a hierarchical fashion, their influence on security. Second, this information is reformulated as a four-layer feed-forward multilayer perceptron. Third, the MLP weights are tuned on the basis of the learning set augmented with the security *margin* type of information to enhance classification reliability and transform the *discrete* classification information of the tree into a *continuous* security margin.

Among the possible ways to reformulate a DT as an equivalent neural network, we have tentatively used the one proposed in [Set90]. It consists of the following four-layer structure [WA93].

1. The INPUT Layer (IL) contains one neuron per attribute selected and tested by the DT. Their activation levels correspond to the attribute values of the presented state.

2. The TEST layer (TL) contains one neuron per DT test node. Each TL neuron is linked to the IL neuron corresponding to the tested attribute.

3. The ANDing layer (AL) contains one neuron per DT terminal node. Each AL neuron is connected to the TL neurons corresponding to the test nodes located on the path from the top node towards the terminal node. Its activation level is high only if the state is directed to the corresponding terminal node of the DT.

4. The ORing layer (OL) contains one neuron per DT class, connected to the AL neurons corresponding to the DT terminal nodes where the class is the majority class. Its activation is high, if at least one of the AL neurons is active.

In order to replicate exactly the classification of the DT, sharp activation functions must be used, to make the transition from -1 to 1 sufficiently sharp, when a state crosses the security boundary defined by the DT.

If the neural network is used to approximate a continuous security margin, rather than to merely classify, some modifications are required. First, the output layer would be replaced by a single output neuron, fully connected to all neurons of the AL. In addition, since the weights as given by the DT translation are not necessarily appropriate, it relies on learning to choose them correctly. To obtain a smooth, easily adaptable input/output mapping a rather smooth transition function is used.

However, in order to obtain meaningful results, and in particular to avoid over-fitting problems, it is important to take care about the normalization and truncation of the margin before the back-propagation algorithm is used to adapt the weights of the ANN. This is because the attributes used to formulate the decision tree may not be sufficiently informative to determine the margin when the latter is much smaller or much larger than the threshold used to define the security boundary. Thus this kind of approximate margin information will essentially be valid only *locally* around the security boundary.

Features. As compared to "pure" MLPs, the method allows one to significantly reduce the time required during the training stage. In particular, the structure of the MLP is directly derived from the decision tree whose complexity is tailored to the problem at hand and the amount of information available in the LS. This means that the tedious task of network structure optimization is avoided. Thus, all in all, network

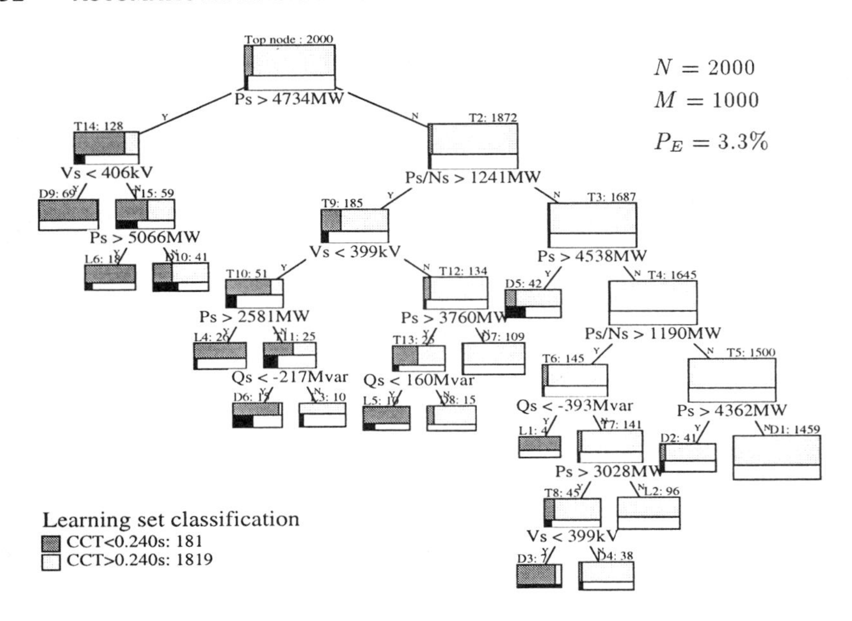

Figure 6.2. Decision tree for transient stability assessment

training times may be reduced significantly (more than a factor 10 in practice) while still being able to improve the accuracy with respect to decision trees.

Furthermore, the MLP model being simpler and close to the original decision tree, its black box nature is somewhat reduced with respect to pure MLPs.

Illustration. Let us briefly illustrate the DT-ANN approach on a transient stability data base relative to the EDF system (see §10.2.2) described in [WA93]. The data base comprises 3000 operating points, and we use 17 candidate atttributes. The output information is a two-class classification with respect to a three-phase short-circuit. We use the first 2000 states as a learning set and the remaining 1000 as a test set.

A DT was built for the above classification; it is represented in Fig. 6.2. We note that the DT consists of 15 test and 16 terminal nodes, and has automatically selected only attributes related to the study plant (Ps, Ps/Ns, Qs and Vs) among the 17 candidate ones. Its error rate determined on the test set $\boldsymbol{TS}$ amounts to 3.3%.

The DT was translated into an MLP composed of 4 input, 15 test, 16 anding and 2 output (classification) neurons, containing 138 parameters, shown in Fig. 6.3. The weights of the latter were updated using the BFGS method together with the learning set. At convergence the error rate was evualated on the $\boldsymbol{TS}$, yielding a slightly improved error rate of 2.7%.

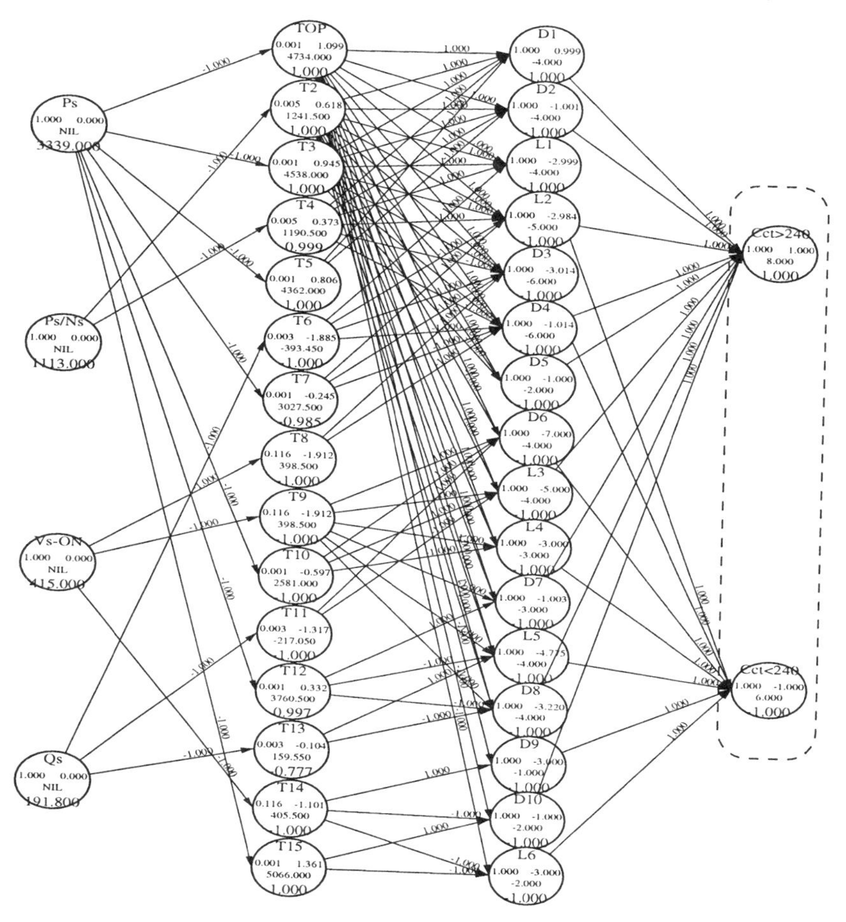

Figure 6.3. MLP resulting from the translation of the DT of Fig. 6.2

Furthermore, in order to exploit the critical clearing time information by the MLP, we merged its two output neurons in a single one, and restarted the BFGS to the learn the CCT values, within a restricted interval around the classification threshold. After convergence we used the resulting network to predict CCT values of the test states and classify them accordingly. This resulted in a further (significant) reduction of the error rate to 1.5%. Notice that this result is very close to the best results obtained on this data base. Actually, only the fuzzy decision tree illustrated in Fig. 5.17 and the hybrid DT-KNN-GA method described below obtained slightly better results (resp. 1.3% and

1.2%). Notice also that this way of combining decision trees with MLPs allows to reduce MLP training times significantly.

6.3.2 Decision trees and nearest neighbor

In Chapter 1 we mentioned that the pure nearest neighbor method is highly sensitive to irrelevant or redundant attributes and its performance may quickly decrease in high dimensional attribute spaces. The only way of coping with this curse of dimensionality consists of adapting the distance to the problem specifics at hand, i.e. augment the nearest neighbor technique with an ad hoc supervised distance learning algorithm. In addition, in the KNN method the appropriate value of K, the number of nearest neighbors should also be adapted.

The objective of the distance learning algorithm can be formulated as follows :

> *Given a LS of objects and a general distance measure defined in the attribute space (see §2.6.1) define the parameters of the distance in order to maximize the reliability of the KNN method in the LS.*

The parameters of the distance are the weights w_a for numerical attributes and the distance tables $\delta_a(v_i, v_j)$ for symbolic attributes. These parameters should thus be adjusted in order to tune the distance to the problem at hand. In particular, those attributes which are less relevant should use small values of weights (at the extreme, irrelevant attributes should correspond to zero weights), while the highly discriminant attributes would correspond to larger values.

Note that in order to assess performance on the $\boldsymbol{LS}$ the leave-one-out method should be used in order to avoid over-fitting.

Whether the actual problem is one of classification or of regression the above distance optimization problem is rather tricky. Indeed, the quality of a nearest neighbor method varies in piecewise constant or at least non-differentiable way when the weights of the distance are changed. In addition, in practical situations where there are often significant correlations among some attributes it can be expected to have several local minima. Thus, the use of classical optimization techniques such as the gradient descent or the BFGS used for MLP learning are precluded. Further, exhaustive search is not feasible as soon as the number of attributes is larger than say 5, i.e. in most of the situations where the method would be needed.

Below we discuss two different heuristic techniques which may be combined in order to adapt the distance in an automatic way. For the sake of simplicity we restrict our presentation to the case of classification problems and numerical attributes only.

Distance derived as a byproduct of tree induction. This is a direct approach which derives the distance from the information provided by building a decision tree for the same problem.

In §5.3.6 we explained an interesting information provided as a by-product of tree induction is the contribution of the different candidate attributes in the information quantity of the tree. From this information, the weights in the distance may be defined in the following way [HWP95, HWP97a, HWP97b]

$$w_a = \frac{\%I_a^T(\boldsymbol{LS})}{\sigma_a(\boldsymbol{LS})} \tag{6.15}$$

where $\%I_a^T(\boldsymbol{LS})$ denotes the percentage of information provided by $a(\cdot)$ in the tree$T$ and the $\sigma_a(\boldsymbol{LS})$ its standard deviation estimated in the learning set.

Once the distance is defined, the optimal value of K in the KNN method may be determined by exhaustive search, e.g. by comparing the results obtained by the leave-one-out method for values in a reasonable range (say $k \in [1, 2, \ldots 15]$).

While this DT-KNN method is heuristic in nature, in many practical applications it provides a very good guess to a distance measure capturing the important problem specifics. Its main attractiveness is computational, especially in comparison to the more involved approach discussed below.

Distance optimization by genetic algorithms. Genetic algorithms (GAs) have been proposed recently to solve various of the hard optimization tasks encountered in automatic learning (global decision tree optimization, MLP network architecture optimization, decision rule optimization...).

In the broad sense, GAs denotes a large variety of optimization methods based upon the principles of natural evolution theory. In this book we are not going to describe them; thus we kindly refer the reader to the specialized literature (e.g. [Gol89, KD91]). All we need to say here that it is a heuristic optimization method based on random search, and that it is able to handle multi-modal non-differentiable problems in principle.

It can thus be applied to the optimization of distances in the nearest neighbor technique. Refs. [HWP95, HWP97a, HWP97b] describe in more detail this KNN-GA method and its performances in comparison with the above DT-KNN one. It is found that the KNN-GA method is in practice a factor of 10 to 100 times slower, and not always able to outperform the DT-KNN method in terms of accuracy.

Combining DTs and GAs with KNN. From the experimental investigations presented in [HWP95, HWP97a, HWP97b] one can conclude that the best way is to combine these three methods, at least from the accuracy point of view.

In this approach DTs are used to provide a first guess to the distance optimization which is provided as a hint to the GA. Furthermore, the partition induced by the decision tree on the attribute space is used to define regions within which the distance may be optimized locally. This provides a further degree of freedom to the method, which may pay in some difficult problems.

While more work is required to mature this approach it has shown much promise in the applications considered so far. Its main drawback is computational inefficiency, since distance optimization for real large scale problems may take several hours to days on high end workstations.

Thus, while the hybrid DT-KNN method is straightforward, neither KNN-GA, nor this DT-KNN-GA method are practical for the time being for real large scale problems.

6.3.3 Discussion

The preceding chapters have shown the complementary nature of the different automatic learning methods. One general characteristic of many real life power system problems is the high number of possible candidate attributes. Thus, dimensionality problems appear which have to be tackled.

As we have seen, among the automatic learning methods discussed, decision tree induction is presently the one which is able to handle large-scale problems most efficiently. Multilayer perceptrons become excessively slow while nearest neighbor lacks accuracy. Nevertheless, it is possible to use these methods, if they are appropriately modified. In particular, they can be coupled with decision trees in hybrid approaches in order to reduce their weaknesses to some extent.

Table 6.1 taken from [WP96] summarizes the main characteristics of various pure and hybrid techniques, in particular in terms of the type of information they can exploit/provide, their expected level of accuracy, and their flexibility.

In terms of accuracy, there exists no universal panacea. However, while in general each method has its own field of competence, in security assessment problems those methods which exploit margins rather than classes (especially with smooth models) generally provide increased accuracy, and also more refined security assessment.

In particular, the proper way to exploit margins consists of saturating them outside a small window around the relevant classification threshold and building an approximate regression model, using only those attributes which influence the margin value within this window. If the end result searched is a discrete classification, it can be derived straightforwardly by discretizing the output of this model. By doing so one succeeds in taking advantage simultaneously of continuous margins and problem simplicity.

In terms of CPU time, the variations are much larger. For example, growing a decision tree can be up to 1000 times faster than optimizing the weights of a multilayer perceptron for the same problem. Thus, while the former method may be used in an interactive trial and error fashion, the latter is hardly practical for large data sets, typically encountered in power system security problems.

Table 6.1. Salient features of AL methods applied to security assessment

	Method	Functionalities	Computational Off-line	Computational On-line
Pure	Crisp DTs	**Good interpretability (global).** Discrete. Good accuracy for simple "localized" problems. Low accuracy for complex, diffuse problems.	Very fast	Very fast
Pure	MLPs	**Good accuracy.** Low interpretability. Possibility for margins and sensitivities.	Very slow	Fast
Pure	KNN	**Good interpretability (local).** Conceptual simplicity.	Very slow	Very slow
Hybrid	Fuzzy DTs	**Good interpretability (global).** Symbolic and continuous. More accurate than crisp trees. Possibility for margins and sensitivities.	Slow	Fast
Hybrid	DT-ANN	Combine features of DTs and MLPs	Slow	Fast
Hybrid	DT-KNN	Combine features of DTs and KNNs	Slow	Slow

II APPLICATION OF AUTOMATIC LEARNING TO SECURITY ASSESSMENT

7 FRAMEWORK FOR APPLYING AUTOMATIC LEARNING TO DYNAMIC SECURITY ASSESSMENT

Security assessment (SA) and in particular *dynamic security assessment* (DSA) is a very versatile topic and is generally approached in a pragmatic, power system and problem-specific way. Although there are many different more or less sophisticated tools, the most widely accepted one is numerical simulation. In planning or operations planning departments, engineers thus use numerical simulation and their own expertise to run some scenarios and extract, by hand, the relevant security information, and finally produce guidelines.

Thus, *the purpose of applying automatic learning to DSA is not to replace, but rather to enhance the existing practice by rendering automatic some of the manual tasks* (selection of scenarios and extraction of relevant synthetic knowledge from the simulation results). The engineers are thereby freed from the most tedious parts and may concentrate on the most interesting ones.

Further, we will see that this approach provides a sound DSA methodology, stating input hypotheses explicitly and extracting reproducible properly validated output results. This enables one to easily repeat a security study with different hypotheses, and adapt the resulting information to changing conditions. It makes also the sharing of information among different people much more straightforward.

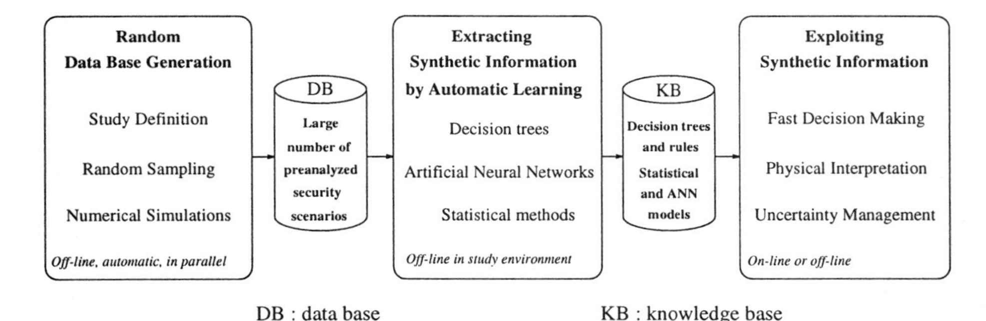

Figure 7.1. Automatic learning framework for security assessment

Thus, in times were human expertise within electric Utilities tends to be threatened, the automatic learning framework provides a means to maintain and even enhance it.

Figure 7.1 depicts the three-step framework to apply automatic learning to DSA.

Random sampling techniques are considered to screen all relevant situations in a given context, while existing numerical simulation tools are exploited - if necessary in parallel - to derive detailed security information.

The heart of the framework is provided by automatic learning methods used to extract and synthesize relevant information and to reformulate it in a suitable way for decision making. This consists of transforming the data base (DB) of case by case numerical simulations into a power system security *knowledge base* (KB). We have seen that, due to their interpretability, decision trees are the cornerstone of the automatic learning tool-box, the other methods being used for their complementary features according to the type of information they may exploit and/or produce.

The final step consists of using the extracted synthetic information (decision trees, rules, statistical or neural network approximators) either in real-time, for fast and effective decision making, or in the off-line study environment, so as to gain new physical insight and to derive better system and/or operation planning strategies.

7.1 DATA BASE GENERATION

The first step of the data base generation consists in specifying the range of scenarios that the particular study will address. While the approach aims at enabling the engineers to carry out more systematic and more global studies, it is generally necessary and even desirable to restrict the focus of a study to an a priori well defined scope.

Starting with the existing expertise and problem statement, random sampling specifications are set up, generally through a sequence of discussions among experts in different fields, such as power system dynamics, protections and economics. The other part of the data base specification concerns the choice of the parameters which will be extracted from the simulations and stored in the data base.

Note that the data base size and the computation time needed to generate it may vary strongly from one application to another, depending on the type of dynamic model used and the scope of the study. However, to be representative the data bases comprise generally a few thousand simulation scenarios, and in typical large scale applications there may be several hundred parameters.

7.2 APPLICATION OF AUTOMATIC LEARNING (DATA MINING)

Application of automatic learning (or data mining) is the most interesting task. The idea is to exploit the capabilities of various statistical tools in order to "discover" interesting information. We have argued that to fully exploit security information data bases it is indeed necessary to combine different automatic learning methods offering complementary features.

At each step of the process the information extracted is validated against an independent test set and against prior expertise and physical knowledge existing about the problem. In particular, the scenarios which lead to dangerous errors may be identified and analyzed in detail. Sometimes, the engineer deems that a new data base should be generated in order to study some situations which were not sufficiently well represented in the original one, or in order to assess the effect of countermeasures suggested by the data mining results.

The final step, as in traditional security studies, is report writing and translating the information extracted from the data base into guidelines for operators and planning engineers.

7.3 EXPLOITING EXTRACTED INFORMATION

Anticipating on the following chapters, let us briefly discuss how this framework may complement classical system theory oriented methods for security assessment.

In practice, there are three dimensions along which we may expect important fallouts.

First of all, let us discuss *computational efficiency*. By using synthetic information extracted by automatic learning, instead of numerical methods, much higher speed may be reached for real-time decision making. Further, in terms of data requirements, whereas analytical and numerical methods require a full description of the system model, the approximate models constructed via automatic learning may be tailored in order to exploit only the significant and/or available input parameters. Computational efficiency was actually the motivation of Dr Dy Liacco, when he first envisioned in the late sixties the use of pattern recognition for on-line DSA [Dy 68]. Even today, this remains a strong motivation.

But the synthetic information extracted by automatic learning methods may itself be complementary to and generally more powerful than that provided in a case by case fashion by existing analytical or numerical methods. In particular, much more attention is paid nowadays to *interpretability* and management of *uncertainties*, the two other important fallouts expected from automatic learning methods.

As concerns *interpretability*, the use of automatic learning to provide physical insight into the nonlinear system behavior was first proposed by Professor Pao and Dr Dy Liacco in the mid-eighties [PDB85]. In the meantime, it has been shown that machine learning (in particular, decision tree induction) is indeed an effective way to extract interpretable security rules from very large bodies of simulated examples [WP93b, WVP+94]. We have already illustrated how extracted rules express explicitly problem specific properties, similarly to human expertise, and hence may be easily appraised, criticized and eventually adopted by engineers in charge of security studies. Moreover, the flexibility of the automatic learning framework allows one to tailor the resulting information to analysis, sensitivity analysis and control applications.

As concerns management of *uncertainties*, we will see that the above framework is indeed able to take into account and manage uncertainties on dynamic models, external systems, load behavior, measurements...

To conclude this introduction, we note that the above motivations for using automatic learning, although presented in the specific context of security assessment, are also valid in many other application fields.

7.4 OUTLINE OF PART II

The aim of Part II is to discuss the large diversity of potential applications of AL to security assessment (although static security assessment can also take advantage of AL, we focus on dynamic security assessment where more needs are felt today).

Chapter 8 discusses security assessment problems, according to the type of physical phenomena involved and the type of practical environments where it is carried out (planning, design, operation...), and indicates where AL can be most useful.

Chapter 9 describes in detail how security information data bases can be generated automatically for real large scale power systems. This is paramount for the success of the approach since the quality of information extracted by automatic learning directly depends on the quality of the security information data base. The ideas are illustrated on two real studies carried out respectively on the Hydro-Québec and the EDF systems.

Chapter 10 provides a representative sample of real-life applications carried out so far for transient stability, voltage security and the study of power system breakdown modes, thereby further illustrating the potentials of AL.

Chapter 11 synthesizes how AL is complementary to existing analytical or numerical techniques in the context of DSA and Chapter 12 bridges the gap between present day practices and probabilistic security assessment, which is likely to become practice in the future. It also discusses the main challenges for research and development.

8 OVERVIEW OF SECURITY PROBLEMS

In this chapter we provide a brief overview of power system security and possible applications of automatic learning.

We start by reviewing the different types of physical problems, restricting our focus on DSA. Then we consider the different working environments where security assessment tools are needed, and comment on the applicability of automatic learning.

8.1 OPERATING MODES

Security assessment consists of evaluating the ability of the power system to face various disturbances and of proposing appropriate remedial actions able to counter its main weaknesses, whenever deemed necessary. Disturbances may be due to external or internal events (e.g. faults subsequent to lightning vs operator initiated switching sequences) and may be small (slow) or large (fast) (e.g. random behavior of the demand pattern vs generator or line tripping).

The different operating modes of a power system were defined by Dr Dy Liacco [Dy 68]. Figure 8.1 shows a more detailed description of the "Dy Liacco state diagram".

Preventive security assessment is concerned with the question whether a system in its normal state is able to withstand every plausible disturbance, and if not, preventive

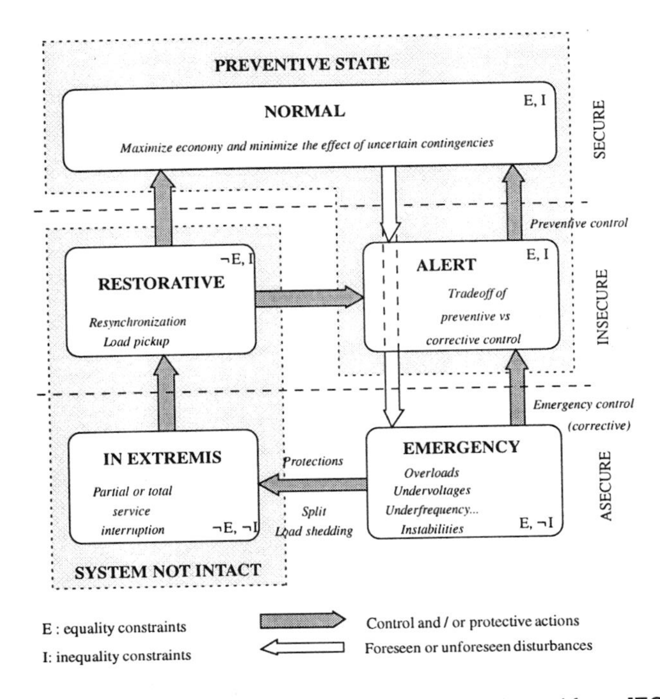

Figure 8.1. Operating states and transitions. Adapted from [FC78]

control would consist of moving this system state into a secure operating region. Since predicting future disturbances is difficult, preventive security assessment will essentially aim at balancing the reduction of the *probability* of losing integrity with the economic cost of operation.

Emergency state detection aims at assessing whether the system is in the process of losing integrity, following an actual disturbance inception. This is a more deterministic evolution, where response time is critical while economic considerations become temporarily secondary. Emergency control aims at taking fast last resort actions, to avoid partial or complete service interruption.

When both preventive and emergency controls have failed to bring system parameters back within their inequality constraints, automatic local protective devices will act so as to preserve power system components operating under unacceptable conditions from undergoing irrevocable damages. This leads to further disturbances, which may result in system splitting and partial or complete blackouts.

Consequently, the system enters in the *restorative* mode, where the operator's task is to minimize the amount of undelivered energy by re-synchronizing lost generation as soon as possible and picking up the disconnected load, in order of priority.

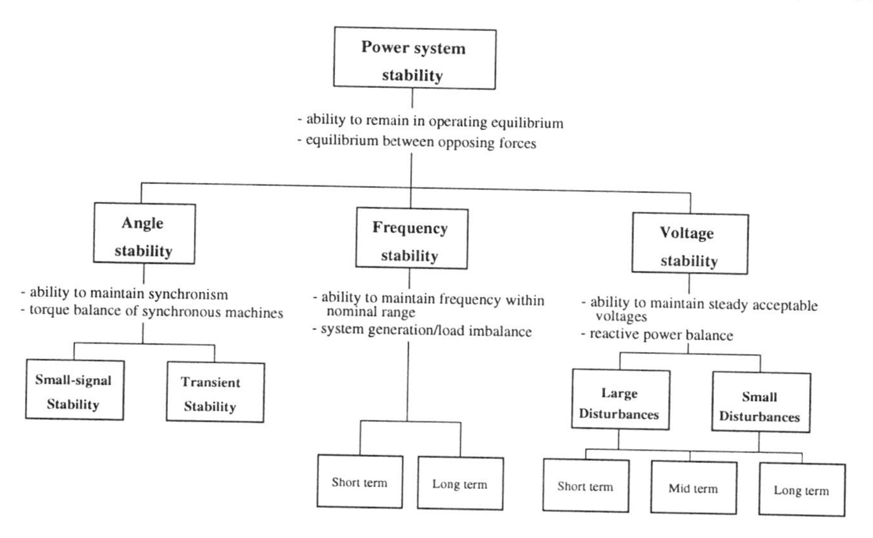

Figure 8.2. Types of power system stability phenomena. Adapted from [KM97]

While automatic learning may be useful in the context of restoration [KAG96], we restrict our discussion to preventive and emergency modes.

8.2 PHYSICAL CLASSIFICATION OF DSA PROBLEMS

Below, we make a very short discussion of the main physical aspects of DSA. We refer the interested reader to [SP97, Kun94] for further reading.

Figure 8.2 taken from [KM97] provides an overview of various types of stability problems which have to be tackled in DSA. Beneath each type of stability problem is indicated the type of phenomena which are characteristic of this type of instability and the type of physical causes which drive them.

Various security problems are distinguished according to the characteristic symptoms (low voltage, large angular deviations...), and the control means (reactive power, switching...) to alleviate problems, the time scales of the dynamics, and further the amplitude of disturbances.

8.2.1 Transient stability

Transient stability concerns the ability of the generators of a power system to recover synchronous operation following the electro-mechanical oscillations caused by a *large* disturbance.

In this context, the dynamic performance is a matter of seconds (if not factions of seconds) and is mainly affected by switching operations and fast power controls (e.g. fast valving, high voltage direct current converters, flexible alternating current transmission systems (FACTS)) and voltage support by the automatic voltage regulators of synchronous generators and static VAr compensators (SVCs).

To determine the *degree* of stability we may evaluate the critical clearing time of a fault, which is the maximum time duration it may take to clear the fault without the system losing its ability to maintain synchronism. We may also evaluate transient energy margins using direct methods [PM93].

8.2.2 Voltage stability

Voltage stability concerns the ability of a power system to maintain acceptable voltages at all nodes under normal and contingent conditions. A power system is said to have entered in a state of voltage instability when it experiences a progressive, uncontrollable decline in voltage [Tay93].

In the context of voltage stability, the fastest phenomena are characterized by sudden voltage collapses developing at even higher speeds than loss of synchronism. More classical is the *mid-term* voltage instability, which corresponds to a typical time frame of one to five minutes. In this case voltage collapse is mainly driven by automatic transformer on-load tap changers trying to restore voltage nearby the loads. There is a third, even slower time frame, corresponding to the so-called *long-term* voltage instability, which involves the gradual buildup in load demand. This interacts with classical static security and is quite within the scope of operator intervention.

Note that in some particular contexts these phenomena may interact strongly and their distinction becomes useless.

8.2.3 Static security

If the transient (voltage and angle) phenomena following the inception of a disturbance are stable, the system will reach an equilibrium point within a few minutes. Static security concerns the viability of this latter, and considers whether voltages and line currents are with normal bounds. If this is not the case, some protections would trigger and the system could experience cascading outages.

Although Static security is by definition out of the scope of DSA, it may interact with dynamic security. In particular, it may happen that an initially stable system is driven out of its dynamic security region through cascading outages of overloaded lines. Thus, static insecurity may be an initiating factor leading subsequently to loss of synchronism or voltage collapse phenomena. For example, the study described in §§9.6.2 and 10.4 considers blackout scenarios where both phenomena coexist and interact.

Nevertheless, since the so-called static phenomena span over significantly longer periods of time (20 to 30 minutes), operator based emergency control may be possible within the context of static security, provided that corrective actions have been prepared in the preventive mode [HP91].

In all other cases, emergency control must be carried out automatically.

8.3 PRACTICAL APPLICATION ENVIRONMENTS AND POSSIBLE USES OF AL

Table 8.1 shows the practical study contexts or environments which may be distinguished in security assessment applications. The second column specifies how long in advance (with respect to real-time) studies may be carried out; the last two columns indicate respectively if an operator is involved in the decision making procedure and if an expert in the field of power system security is available.

Below we screen these contexts and discuss useful applications of automatic learning. By doing so we distinguish between the context where data bases are generated and automatic learning is applied, on the one hand, and the context where the results of this work may be applied, on the other. Thus, while the results of automatic learning may be used in any context, we believe that, at the present stage of research, the task of building data bases and applying automatic learning are best carried out in off-line study environments under the control of experts in power system security assessment. Only after the methodology will be fully accepted and understood in the off-line environments, it may be possible to carry data base generation and automatic learning to the on-line environment of control rooms.

8.3.1 Generation-transmission system planning

In system planning, multitudinous configurations must be screened for several load patterns, and for each one a large number of contingencies. An order of magnitude of 100,000 different scenarios per study would be realistic for a medium sized system. While enough time may be available to carry out so many security simulations, there is still room for improved data analysis methods to exploit their results more effectively for the identification of structural system weaknesses and to provide guidelines to improve reliability.

Note that the probabilistic Monte-Carlo simulation[1] based planning tools could also be adapted so as to take advantage of automatic learning methods. For example the scenarios generated by random sampling could be stored in a security information data base and further analyzed by automatic learning. The results of automatic learning could then be reused so as to define better sampling schemes in order to reduce variance in subsequent studies [Weh95a].

Table 8.1. Security assessment environments. Adapted from [WP93a]

Environm.	Time scales	Problems	Operator	Expert
System planning and design	1 - 10 years	Generation-Transmission system Protection systems Control systems	No	Yes
Operation planning	1 week - 1 year	Maintenance Unit commitment Protection settings	No	Yes
On-line operation	1 hour - 1 day	Preventive mode Security assessment	Yes	Partly
Real-time*	sec. - min. - hour	Emergency control Protective actions	No**	No
Training	months - days	Improve operator skill	Yes	No

* Here we distinguish between *real-time*, which considers dynamic situations following a disturbance inception, from merely *on-line* which considers static pre-disturbance situations.
** except for static security corrective control

8.3.2 Design of protection and control systems

The other type of task which is carried out in the off-line study environments concerns the design and validation of all kinds of protection and control systems. We will elaborate a little more on this topic, since automatic learning could be particularly useful in this context. For the sake of clarity we will not distinguish between the types of protection and control systems considered (local, centralized, special stability controls, . . .).

As concerns existing protection and control systems, the automatic learning framework would be useful in terms of analysis, in order to evaluate performances in the context of diversified simulation scenarios. Furthermore, the automatic learning methods could be used so as to tune various parameters (gains, temporizations, thresholds. . .) in the most effective way and find out strategies to decide how to adapt these latter to changing operating conditions (e.g. winter and summer settings of low voltage thresholds for automatic tap changer blocking schemes. . .).

As concerns the design of new systems, the automatic learning methods may be used in order to identify the most appropriate real-time measurements and/or signals and to determine appropriate control laws or protection logics. The resulting systems may then be validated against a diversified test set of simulation scenarios before going toward field tests.

Note that, while the distinction of different types of phenomena provided in Fig. 8.2 is interesting from a conceptual point of view, it may become irrelevant in the context of the most extreme operating modes when the power system is undergoing a breakdown scenario.

Thus, in the context of special stability control systems' design it may be wiser to look at the power system dynamics as a single global phenomenon, in order to be able to study interactions of phenomena and related protection systems. In Chapter 9, we consider an example data base generation within this context, drawn from a research collaboration between Electricité de France and the University of Liège [WLTB97a].

8.3.3 Operation planning

Operation planning, as suggested in Table 8.1, concerns a broad range of problems, including maintenance scheduling (one year to one month ahead), design of operating strategies for usual and abnormal situations, and setting of protection delays and thresholds. The number of combinations of situations which must be considered for maintenance scheduling is also generally very large, and automatic learning approaches will be useful to make better use of the available information and to exploit the system more economically.

In the context of operation planning, it may be possible to re-tune protection and control law parameters in order to adapt them to unforeseen conditions, yielding similar applications of the automatic learning framework than those discussed above in the context of the design environment.

Similarly, for the closer to real-time determination of operating security criteria it would allow engineers to screen more systematically representative samples of situations, in order to identify critical operating parameters and determine their security limit tables needed for on-line operation.

These types of applications are further illustrated in this book in many places (e.g. §§9.6.1, 10.2.2, 10.3, and 11.1).

8.3.4 On-line operation

In on-line operation, it is presently not feasible to generate data bases automatically nor to extract information from them by applying automatic learning.

However, the data bases generated off-line and decision rules extracted from them may be exploited for on-line decision making. On-line operation will also provide the required feedback to the engineers in charge of defining strategies, when major changes happen in the system.

In the future, with faster computers and more efficient security assessment tools it is also conceivable that data bases and security criteria might be refreshed automatically on-line [DL97] . We mention also the very ambitious proposal made by Dr Rovnyak and Prof. Thorp from Cornell University which aims at building on-line decision trees for real-time stability control [RKTB94] .

8.3.5 Real-time monitoring and control

A fortiori, in the context of real-time monitoring and control, today it is not feasible to build data bases and apply automatic learning.

However, as we mentioned above, automatic learning may be used off-line to design criteria to trigger automatically emergency control actions, so as to prevent a disturbed system state to evolve towards blackout.

Even more than in the preventive mode studies, it is important to use appropriate models to reflect the *disturbed* power system behavior, when designing these security criteria, and in particular, to take into account modeling uncertainties and measurement errors while generating the data bases.

Furthermore, the use of readily available system *measurements* as inputs to the derived emergency control rules is often an operational constraint in addition to minimal data requirements and ultra high speed.

From the automatic learning point of view, one of the main difficulties is to handle the dynamic time-varying nature of attributes. Indeed, real-time controls and protection systems must operate in situations where the system is in some dynamic process, i.e. when the measurements used as inputs are changing with time. Thus, it is necessary to take into account the temporal aspects of the dynamics in the process of designing such systems. In automatic learning this translates into attributes whose values are functions of time (see §10.4).

8.3.6 Operator training

During operator training, the security criteria derived in either of the preceding contexts might be usefully exploited as guidelines, provided that they are presented in an intelligible way. In addition, these models might be used internally in a training simulator software, in order to set up particular scenarios presenting particular insecurity modes.

8.4 ANALYTICAL TOOLS

In addition to standard time domain numerical simulation, a rather large set of numerical methods are available for security assessment in the different time frames mentioned [CM97] . We call them *analytical* tools since they exploit analytical power system models in contrast to the *synthetic* ones extracted by automatic learning techniques.

All these tools, may be exploited during the data base generation, provided that they are accepted by the concerned Utility.

Furthermore, the automatic learning framework may be used in order to assess simplified tools and/or simplified dynamic models, by comparing systematically and on a representative sample of simulation scenarios their security assessment with the one provided by reference methods and/or reference dynamic models. This allows indeed to validate simplified models and calibrate security indicators provided by fast screening tools. Such applications have already been carried out as a byproduct of some of the security studies described in Chapter 10, in order to validate simplified methods for transient stability and voltage security, and determine their validity and error bounds with respect to detailed numerical simulation.

8.5 SUMMARY

The effect of a contingency on a power system in a given state can be assessed by numerical (e.g. time-domain) simulation of the corresponding scenario or by other analytical tools. However, the nonlinear nature of the physical phenomena and the growing complexity of real-life power systems make security assessment a difficult task. For example, the everyday monitoring of a power system calls for fast analysis, sensitivity analysis (which are the salient parameters driving the phenomena, and to which extent ?), suggestions to control.

Thus, there is a very large diversity of security problems and the way they are tackled in practice is generally power system specific. Very often also, the methodologies, models and criteria used in planning environments are different from those used for operation, which leads to further difficulties.

The automatic learning framework, due to its flexibility, offers promising capabilities in all these problems, and proposes a unified methodology which can be used in all environments. Thereby, information could be shared more easily and more systematically between planners, operation planners and operators. As we pointed out, it enables one also to take into account modeling uncertainties and measurement errors in the security assessment task.

The sceptic reader might wonder whether all this is really feasible, or how well it could work in practice for complex large scale power systems. The next chapters of this book will provide some answers.

In particular, in the next chapter we will describe a sound methodology and technical means to generate high quality security information data bases, in order to make these capabilities become reality. We will also illustrate it on some real-life case studies. However, in order to appraise actual interest in practice we will have to wait until Chapter 10, where a great deal of the theory will be illustrated on real case studies.

Notes

1. Monte-Carlo simulations are nowadays used in planning studies in order to estimate expectations quantities such as operating costs and load-curtailment for future operating conditions. In its most elementary form, it consists of using random sampling to screen possible scenarios according to their probability, together with numerical methods to compute for each scenario the operating cost and amount of load curtailed, and averaging this values for the simulated scenarios.

9 SECURITY INFORMATION DATA BASES

In this chapter we describe in detail the methodology and tools needed to generate sound data bases. The quality of the security information data base is paramount. If the data base is biased, unrepresentative, or too small, then the information extracted by automatic learning will probably be useless.

In the literature, clever techniques have been proposed to a priori reduce the size of data bases, and hence reduce the computational burden. Indeed, by choosing the simulation scenarios in a sequential fashion, similar to the trial and error procedures used by human experts one can try to localize the scenarios at the vicinity of security boundaries. Unfortunately, the scenarios contained in such data bases are correlated, and strongly biased by prior information. Our experience has shown that this may dangerously affect the quality of the information extracted by automatic learning.

Our point of view is that today computing power and storage are not a bottleneck anymore, and they become cheaper and cheaper everyday. What is really needed, is a sound methodology and appropriate software to take advantage of it. Thus, in order to avoid bias, in our methodology a data base is first specified (study scope, random sampling, extracted information), then the scenarios are generated automatically, and finally they are simulated (if necessary, by exploiting parallel computations). In particular, during random sampling, simulation scenarios are chosen independently from each other, and before automatic learning is applied, the data base goes through a validation stage.

Below, we first start by describing what we mean exactly by a security scenario. Then, we go through the overall process of data base generation, and discuss in detail the different steps.

The methodology has crystallized during research collaborations with industry (Electricité de France and Hydro-Québec) in the context of large scale DSA problems (transient stability, voltage stability, preventive and emergency modes). At the end of this section we will briefly describe two examples of these.

9.1 SECURITY SCENARIOS

In the context of a particular power system and DSA problem, a security scenario is defined by the three following components : initial operating point, external disturbances, dynamic modeling hypothesis. In some security studies all three components may vary randomly from one scenario to another. In other cases some are kept constant. For example, in the simple illustrative example of §1.2, only the operating point was varying. In the case study described in §10.3, all three components vary.

Below we will further comment on each scenario component, in order to highlight different sampling strategies corresponding to different types of studies.

9.1.1 Initial operating point (OP)

The initial operating point is a static equilibrium at which the system is supposed to sit, before any external events start initiating dynamics. It is defined by available equipments in operation such as generators, lines, transformers, SVCs, capacitors, reactors,. . . , substation topologies, load level and pattern, generation schedules, interconnection tie-line flows, and voltage/VAr dispatch.

Depending on the kind of study, it may be a normal secure or insecure state, optimally dispatched or not, viable or not. For example, in preventive security assessment studies we can consider all kinds of random viable states, in order to find out differences between secure and insecure ones, independently of any a priori operating philosophy, since the purpose is precisely to find out such philosophies. In other studies, for example considering the design of emergency controls, it may be interesting to consider only a small number of normal, N-1 secure operating states (see below).

9.1.2 External disturbance (ED)

The external disturbances are the events which will initiate the dynamics and drive the system away from its equilibrium. Depending on the type of study, they may be simple outages, load disturbances, faults, or any kind of combination of these.

For example, presently, in most Utilities the policy for preventive security assessment is deterministic. It consists of assuming a list of contingencies which the system must survive. Thus, in the data base these latter should be simulated for each operating point. However, future operating strategies might switch to probabilistic preventive security assessment, leading to the consideration of multiple disturbances, with different probabilities.

Similarly, when designing special stability emergency control systems or defense plans which are supposed to operate only in very unusual, highly disturbed conditions, it would be wiser to consider a much larger diversity of randomized disturbances (e.g. fault duration and location, multiple faults, ...).

9.1.3 Dynamic modeling hypothesis (MH)

The dynamic modeling hypothesis concerns the parameters of the system (generators, lines, loads...) and the assumptions made concerning the behavior of various automatic actions which will take place in the system in order to respond to its dynamics (control loops, protections, special stability controls, reaction of external systems...), as well as manual actions (dispatchers, plant operators) which may interfere in the case of slow dynamics.

Note that, in most classical security assessment studies, the MH is considered to be fixed, just as if it were perfectly known. In the automatic learning approach, it is possible to randomize those aspects which are uncertain, according to the information at hand in the particular study context. For example, in planning studies it would be wise to randomize the characteristics of the not yet installed equipments. In operation planning studies, it might be wise to randomize load models and external system models. In emergency control studies, it might be wise to randomize also relay settings, fault impedances... and take into account possible malfunctioning.

9.2 OVERALL PROCESS OF DATA BASE GENERATION

Figure 9.1 summarizes the three successive steps of the data base generation process.

In practice, the specification is the most important and time consuming stage. As we will see below, it is at this stage that the existing expertise is injected in the automatic learning framework.

The second step is automatic, and may be carried out by appropriate software tools. However, presently there exists no single software package able to encounter all the needs of the various types of security studies. Thus, in the context of our research we have developed a series of specific tools to encounter the needs of specific types of applications. One of them was used in the study described in §10.3. Below we will describe the outlook of a general tool which is under development.

The third and last step calls for engineering judgment together with data mining tools. Before extracting any security criteria from a data base, it is indeed necessary

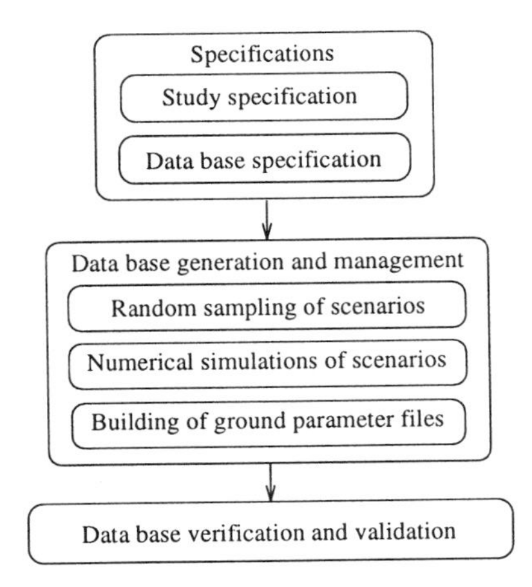

Figure 9.1. Overall data base generation process

to verify its consistency with the initial specifications. Below we will provide some indications on the types of problems which might indeed lead to unrepresentative data bases.

9.3 SPECIFICATIONS

9.3.1 Study scope specification

The first step of the data base generation consists in specifying the range of scenarios that the particular study will address. While the approach aims at enabling the engineers to carry out more systematic and more global studies, it is generally necessary and even desirable to restrict the focus of a study to an a priori well defined scope (see [JWVP95], for a detailed discussion).

Starting with the existing expertise and problem statement, and depending on the kind of information expected from the study (e.g. preventive vs emergency), it is decided which parts of the security scenarios will be variable, which parameters to change for each component, what kind of simulations will be carried out (time scales, analytical security assessment tool, level of modeling), what information should be extracted from them.

Some base cases are selected, and a catalog of variable parameters which are important for the study under consideration is set up, as well as contingency lists and uncertain modeling components.

Constraints among the various parameters may also be defined in order to filter out unrealistic scenarios.

9.3.2 Data base specification

Random sampling specifications. In order to finalize specifications, it is first necessary to choose probability distributions for the random sampling procedure. In practice, one starts with existing statistical information about the variability in real life of the considered parameters. However, using this information directly is not possible in general, because it would not lead to rich enough data bases. In particular, in most cases it would lead to a very small number - if any - of interesting scenarios, among the few thousands which can typically be simulated. Thus, in practice it is necessary to bias probability distributions, for example in order to increase the proportion of stressed operating points, or dangerous faults. This is where the engineering judgment comes into play.

Thus the random sampling specifications are set up, generally through a sequence of discussions among experts in different fields, such as power system dynamics, protections and economic issues.

In the random sampling specification, it is useful to distinguish among *primary* and *secondary* parameters. The former are those upon which the security study is focusing, and in terms of which it is desired to characterize security. The latter are those parameters which are either uncontrollable, or unobservable or uncertain : they are made variable in order to yield robust security information.

Concerning operating point parameters, flexibility will depend on the type of tool used to build consistent operating points. For example, if a simple power flow calculation is used to build operating points, then the parameters must be consistent with the load flow equations : thus the random sampling specifications are formulated in terms of independent power flow input variables.

Notice that the simple Monte Carlo sampling scheme outlined above can be made more sophisticated by taking advantage from structured sampling techniques and optimal experiment design. To the interested reader we recommend ref. [KO95] which discusses this very interesting topic in general and ref. [VWM+97] which proposes an application in the context of a specific security assessment problem.

Extracted ground parameters. The other part of the data base specification concerns the choice of the *ground parameters* which will be extracted from the simulations and stored in the data base. These, and combinations of these will be used later as input and output variables for automatic learning.

Again, the type of parameters and how they are extracted from the simulation results will strongly depend on the type of study. For example, in preventive security assessment the input variables will typically be static operating point parameters (power flows, active and reactive generations, topology information) and the output information will be security margins or classes, defined with respect to a contingency list. In emergency control, the input parameters would rather be dynamic system measurements available in real time (voltages, rotor velocities . . .) and output information would measure severity, e.g. incumbent load and generation loss.

Notice that in large scale DSA problems, there may be thousands of state variables and it is generally not necessary to extract them all, but it is important not to miss interesting ones. Our experience shows that in general a few hundred ground parameters are selected, at the data base specification time. Later on, some of them are found to be useless; others, which may be computed as functions of the ground parameters, can be easily added if required.

Acceptability criteria and filtering. During the specification of the data base it is generally not possible to guarantee that all scenarios will be realistic, reasonable, acceptable or even simulatable.

Thus, one should expect that some of the operating point specifications will lead to non converging power flows, or yield an unrealistic state. Similarly, the dynamic simulation tool may fail to simulate some of the very severe scenarios. Therefore, the last step of the data base specification consists of defining acceptability criteria which will be checked during the data base generation in order to filter only those scenarios which are deemed acceptable.

Number of security scenarios simulated. The number of scenarios is a compromise between two contradicting requirements : the larger the data base, the better for automatic learning, but the larger also the required computational resources.

The minimal number of scenarios required to obtain useful automatic learning results mainly depends on the problem complexity, which is generally not known in advance. Thus, a rule of thumb is a few thousand accepted scenarios after filtering. In order to reach this number it may be necessary to generate many more, depending on random sampling specifications, power system specifics and, of course, acceptability criteria.

How many scenarios can be simulated with acceptable response times will depend on the type of dynamic phenomena that are simulated (e.g. mid-term voltage security scenarios can be simulated more efficiently than transient stability ones), and on the type of information that is computed (e.g. margin calculations will take longer than mere security classifications), and of course on the computing power made available.

With present day workstations, the data base generation typically will take between a few hours and some weeks of CPU time. Data base sizes can range from a few MBytes (typical) to some GBytes (exceptional).

Summary. The quality of the information extracted by automatic learning is conditioned by the quality of the security information data base. Thus, the first time a new problem is considered, data base specification needs to be done very carefully, with as much as possible input from Utility experts. The time required to finalize them

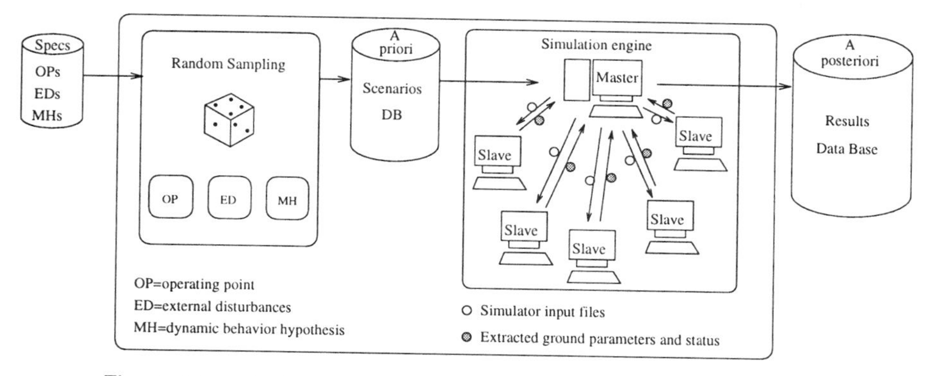

Figure 9.2. Data base generation tool. Adapted from [WLTB97a].

may take a few weeks. Sometimes, a couple of iterations are necessary, involving the generation of small pilot data bases in order to tune various parameters. However, once the required information has been formalized and validated, when subsequently similar problems are considered the adaptation of the random sampling specifications is much more straightforward, and the software developed for the data base generation may be reused easily.

9.4 DATA BASE GENERATION AND MANAGEMENT

We now turn to the computational problems. How to carry out automatically the data base generation, and how to manage the resulting data base ?

9.4.1 Data base generation

The study described in §10.3 uses a sequential data base generation tool which was developed a few years ago in the context of voltage security assessment of the EDF system, and which is now used in real life studies. Here, let us consider the more advanced parallel scheme depicted on Fig. 9.2, very close to the tool used in the study reported in §9.6.2. It is composed of the following modules

1. Random sampling.
   Input : specification of the study scope, in terms of probability distributions, base case data files, dynamic modeling data, number of scenarios to generate, random number seeds.
   Output : a priori data base describing the sampled scenarios.
2. Simulation engine (master/slave organization).
   (a) Simulation input file builder (master).
   Input : a description of a scenario, reference input data files.
   Output : a set of modified input data files.

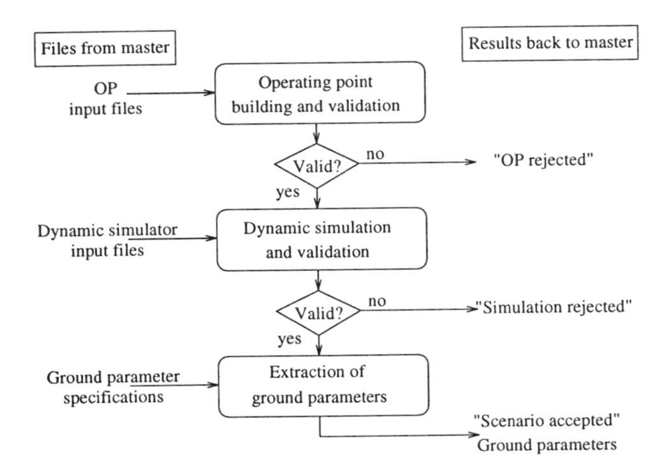

Figure 9.3. Scenario simulation

(b) Task dispatch (master).

(c) Task simulation and extraction of relevant information (slave, see Fig. 9.3).
Input : Input data files, specification of ground parameters to extract.
Output: Extracted attribute files, simulation diagnostics.

(d) Data base builder (master).
Input : simulation diagnostics, results files for each accepted scenario.
Output : results data base.

Master and slaves are standard Unix workstations, exchanging data through files. As soon as a slave becomes idle, it receives from the master a scenario to simulate. Simulation involves three successive steps (see Fig. 9.3) : (i) OP building; (ii) dynamic simulation; (iii) extraction of ground parameters.

A scenario may be rejected at the first stage if the operating point specification is not realistic, or at the second stage if the dynamic simulation fails. Thus each of the dispatched scenarios receives a label : accepted, rejected operating point, rejected simulation. These labels are collected by the master and included in the scenario "a priori" data base, for later validation. For the accepted scenarios, the slave sends also the extracted ground parameters back to the master, who collects this information and puts it into the "a posteriori" security information data base.

9.4.2 Data base management

At the present stage, we found that there is no need to use a sophisticated data base management system. The security studies are merely organized into directories, and the a priori and a posteriori data bases are organized into a series of files. The a priori data base collects random sampling specifications and statuses (accept or reject and reason) of the generated scenarios. The a posteriori data base collects the values taken for the ground parameters of the accepted scenarios. To ease access with standard tools, parameters of the same type are grouped together in flat ASCII files, which are compressed with a standard UNIX compression facility to save space. They may be easily exploited by various automatic learning algorithms, possibly after converting them to the appropriate format.

As we will see below, it is important to keep trace of the scenarios which were rejected so as to be able to analyze the validity of the data base. It is also useful to be able to pick a scenario from the a priori data base and re-simulate it, if required, e.g. for detailed analysis of some interesting scenarios detected during the data base mining.

9.5 DATA BASE VALIDATION

The validation of the data base consists of two steps.

The first step consists of analyzing the scenarios which have been rejected, in particular in order to find out whether the filtering does not alter too strongly the probability distribution of the independent parameters used in the random sampling. This analysis may be carried out by applying the data mining tools, mainly low level statistical visualizations, on the a priori scenario data base. Higher level automatic learning tools, e.g. decision tree induction, may be used in order to find which kind of scenarios are rejected and whether there is some abnormal behavior.

The next step of the data base validation concentrates on the analysis of the a posteriori security information data base. Again, the same types of low level data mining tools are applied in order to check that the information is sufficiently rich, i.e. how the ground parameters are distributed and correlated.

As mentioned above, during validation it is the responsibility of the engineer to decide to accept the data base, and proceed to the next step, or to reject it and suggest modifications to the random sampling specifications in order to generate a new DB.

9.6 EXAMPLES

9.6.1 Hydro-Québec (transient stability)

This study was already introduced in §5.3.1 in order to illustrate decision tree induction. Here we focus on the data base generation and validation tasks.

Study system and data base specification. Within this research, a data base was generated for the Hydro-Québec system corresponding to the situation of summer 1992. The first goal was to screen systematically all relevant "four-link" configurations of the James' Bay corridor, yielding a highly complex set of topologies. The reasons for choosing this situation were the high level of complexity, and the availability of optimized stability limits in LIMSEL (Hydro-Québec's on-line stability limit tables).

In order to generate the data base, the following variables were chosen as parameters of the random sampling procedure.

The power flows in the three important corridors of the Hydro-Québec system are drawn independently in the intervals indicated in Fig. 5.6. The James' Bay corridor corresponds to the study region whereas the Manic-Québec and Churchill Falls corridors are outside the study region but may influence the value of its stability limits.

The generation of the main complexes of hydro-electric power plants are adjusted so as to obtain the chosen power flows, while the distribution among the individual Lagrande and Manic/Outardes plants are randomized to yield a wide diversity among the power flows of the individual lines.

The topology is chosen independently according to a pre-defined list of possible combinations of line outages with respect to the complete five-link topology. Only the James' Bay corridor is modified and only so-called four link topologies are generated. This yields a total of more than 300 possible topologies.

The voltage support devices (SVCs and synchronous condensers) available in the six substations of the James' Bay corridor, indicated in Fig. 5.6, are widely variable during the random sampling since their influence on the stability limits is very strong. Their total number is drawn between 0 and 12 according to predefined probabilities, and their distribution in the substations is randomized.

Data base generation. The above specifications led to the development of a specific software to generate a data base of random operating points. We expected difficulties with load flow convergence. Indeed, the very long distances between remote generation sites and load centers and the longitudinal grid lead to voltage control problems. In particular, the important variation of the power flows in the random sampling induces highly variable reactive losses and hence voltage drops, which may prevent the load flow computation from converging properly, thus leading to a low rate of accepted operating points.

Further, to represent normal operating conditions the reactive compensation needs to be adapted automatically to the power flows. This means switching shunt reactors in the UHV system and capacitor banks on lower voltage levels. Thus, an *automatic reactive compensation* loop was developed by Hydro-Québec and included into the RP600 load flow program used for this study.

In spite of this improvement, the first random samplings yielded a very high percentage (up to 70%) of diverging load flow computations. To be able to analyze the

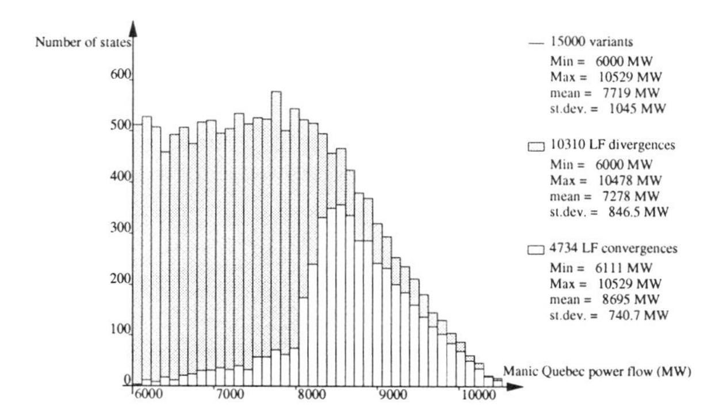

Figure 9.4. Convergence diagram of Manic-Québec power flow (6 base case files)

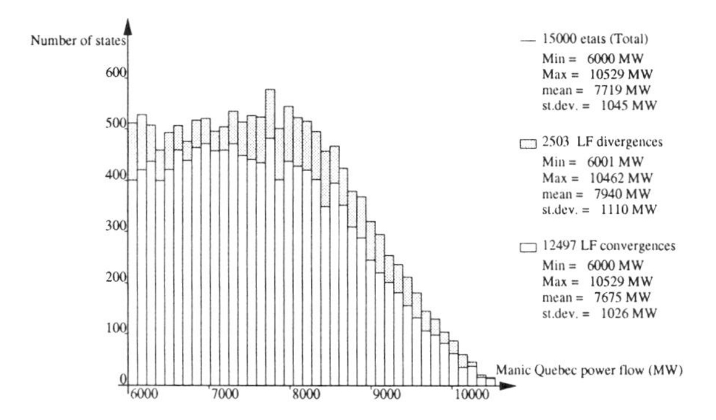

Figure 9.5. Convergence diagram of Manic-Québec power flow (12 base case files)

physical or algorithmic reasons for such high divergence ratios, various frequency diagrams were drawn for the a priori data bases, corresponding to the specifications of the randomly selected variants, classified as *diverging vs converging*.

For example, Fig. 9.4 shows a typical histogram, similar to those obtained in the first a priori data base obtained. The proportion of converging and diverging load flow computations is represented in terms of the specified values of the power flow in the Manic-Québec corridor. One can see that only a small proportion of states did actually converge, and it appears clearly from the diagram that the cases of divergence predominate mainly for power flows below 8,000MW. The reason is that the base case solutions used to initialize the load flow computation were too far away from the solution.

Several iterations were required in order to obtain a satisfactory data base. To improve the convergence we have used a larger number of base cases and a heuristic strategy to choose the appropriate one for each operating point specification. Figure 9.5 reproduces the final distribution of the cases of load flow divergence in terms of the Manic-Québec power flow. With respect to the diagram of Fig. 9.4, one can observe that the proportion of divergences is strongly reduced and they are more or less uniformly distributed. Thus the filtering does not affect the representativity of this attribute in the data base. Similar analyses of the other attributes have shown that the data base was indeed acceptable.

The generation of the of the 12,500 operating points of the final data base, took about one week using 30% of the CPU of a Sun Sparc 10 workstation.

The operating points were then sent back to Montréal and fed into the LIMSEL data base in order to extract stability limits and classify them as stable or unstable; this information was put together with the ground parameters extracted by the load-flow to yield the final data base. The total amount of data (including the a priori data base) was about 20Mbytes, once compressed. Note that, at the time of the project and with the computing power available, using the time-domain simulation program to analyze the stability would have taken several months. We deemed that in order to assess the methodology, the information contained in the LIMSEL data base was sufficiently accurate. As an anecdote, we mention that during data base validation we detected 30 states for which LIMSEL provided erroneous limit values. It was found out to be due to a transcription error in the LIMSEL limit tables which was corrected by Hydro-Québec.

Below we will comment briefly on some automatic learning results extracted from this large data base. Further results are provided in [WHP95a, WHP+95b].

Automatic learning results. The tree partially represented in the right hand part of Fig. 5.13 was built on the basis of the first 10,000 states of the data base and 87 candidate attributes (power flows and generations, topology indicators, VAr support), including four linear combination attributes. All in all, it comprises 57 test nodes and 58 terminal ones. It has identified among the candidate attributes the 24 most relevant ones. Among others, at several test nodes (including the topnode) it has selected a linear combination of the total power flow "Trbj" in the James' Bay corridor and the number of SVCs in operation "Nb_Comp" which thus confirms prior knowledge. Thus, the threshold values of "Trbj" are functions of "Nb_Comp". For example, if "Nb_Comp"=12, the leftmost terminal node "L1" in Fig. 5.13 corresponds to a limit value of

$$\max\{6271 + 12 * 120; 5656 + 12 * 215; 5533 + 12 * 269\} = 8{,}761\text{MW},$$

above which a state with 12 SVCs in operation is unconditionally declared unstable, meaning that there is at least one line-fault in the James'-Bay corridor which would lead to loss of synchronism.

Table 9.1. KNN results for the Hydro-Québec system

K	1	3	5	7	9
67 candidate attributes					
$P_e\%$ (TS)	12.58	11.33	10.53	10.21	10.25
24 attributes of DT of Fig. 5.13					
$P_e\%$ (TS)	6.93	6.73	6.13	6.13	6.61

To evaluate its generalization capability, the tree was tested on the basis of the 2,500 states of the data base not used for its building, yielding an overall error rate of 4.3%. Out of the 1,622 fairly unstable states, only 30 are classified as stable yielding 1.85% "dangerous" errors. On the other hand, 23 marginally unstable states are classified stable, leading to small non-detection errors. There are also 52 false alarms, i.e. stable test states classified unstable by the tree.

To improve accuracy, the same data base was further exploited by building a multilayer perceptron (with a single hidden layer of 20 neurons) on the basis of the same 10,000 learning states. Note that in this case we do not use a security margin as output, no such information being available. Thus the output information of the MLP is in the form of a 0/1 encoding of the security class. At convergence, the MLP yields a reduced test set error rate of 2.4%.

In terms of computational requirements we mention the following CPU times determined on a SUN Sparc10 workstation : 1 hour for the decision tree building and 1 second for testing the 2,500 test states; 60 hours for the learning of the MLP weights and 10 seconds for testing it.

Table 9.1 shows the accuracy results obtained with the KNN classifier for two different cases. The first line of results corresponds to the use of all 67 attributes in the distance computation[1]. The results are quite disappointing with respect to the decision tree and the multilayer perceptron. We note that the value of $K = 7$ provides the best results. The second line of results corresponds to using only the attributes identified by the decision tree of Fig. 5.13 : the reliabilities are significantly improved with respect to the preceding ones but the level of performance of the best DTs or MLPs is not reached; here again the value of $K = 7$ yields the best results. The well-known high sensitivity of the nearest neighbor to the attributes used in the distance computation (and more generally to the weights used in the distance) is observed here very clearly.

Other AL results obtained on this data base with various automatic learning methods are summarized in [WHP+95b, WHP95a, HWP97b].

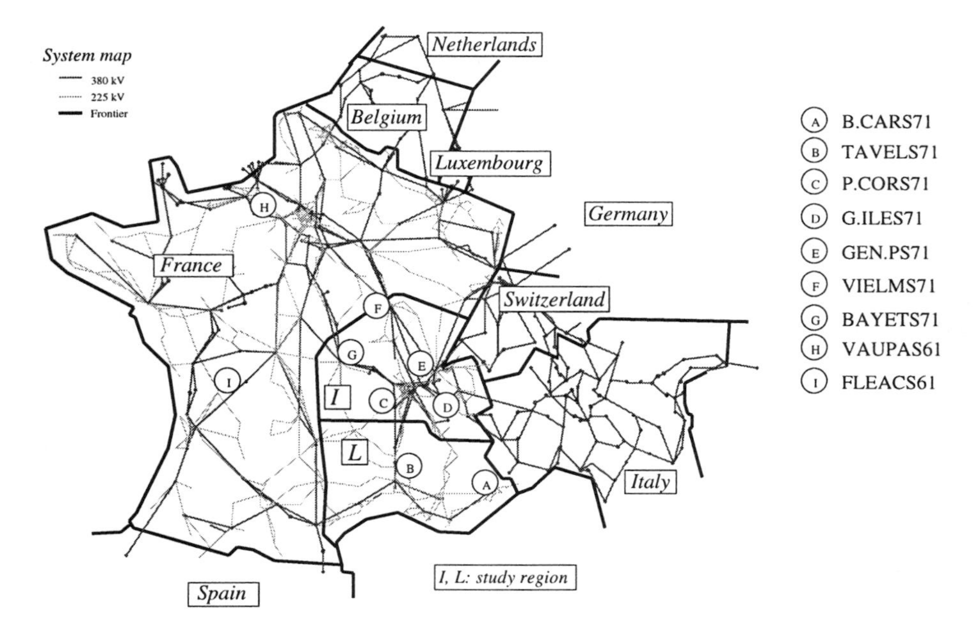

Figure 9.6. One-line diagram of the study region and surroundings

9.6.2 Electricité de France (breakdown scenarios)

In order to further illustrate the diversity of problems to which the automatic learning framework may be applied, let us briefly describe the data base generated within a recent research project [WLTB97a]. The long term goal of this research is to develop a global probabilistic approach to analyze and improve the dynamic performance of power systems in extremely disturbed modes, i.e. under circumstances where various fast and slow dynamic phenomena and their corresponding special protection schemes tend to interact yielding cascades of very intricate behaviors.

Specifications. In July 1995 a research collaboration was started between the University of Liège and Electricité de France, to apply the automatic learning framework to a case study on the EDF system. EDF experts defined a study region in the Provence/Alpes/Riviera subsystem, which was already known to present rather diverse failure modes : cascading line trippings, plant and area mode loss of synchronism, voltage collapse.

This South-Eastern part of the EDF system (see Fig. 9.6) is generally exporting large amounts of power to the rest of France and towards foreign countries (Italy, Switzerland and indirectly Spain). In the very extreme South-Eastern part it is weakly meshed and deficient in generation, thus liable to voltage collapse phenomena. This subsystem is already equipped with various automatic emergency control systems, in order to mitigate various types of failure modes.

It was decided to focus the study on the effect of multiple disturbances and abnormal operation of protection systems, which are often the causes of power system failures. On the other hand, in order to reduce the amount of software developments for the data base generation, it was decided to choose three operating points in a manual fashion. Then the random sampling specifications were set up for the EDs and MHs.

To enable the simulation of both fast and slow phenomena, while taking into account the operation of the relevant protections and special stability control systems, a rather detailed dynamic model was first set up. This model, comprising all in all more than 11,000 state variables is described in [WLTB97a] .

In order to be able to analyze in detail different modes of failure of the power system, it was decided to build up a data base containing temporal attributes, i.e. curves representing the variation in time of various quantities deemed relevant (see below). A specific curve interpolation routine was developed in order to extract these curves from the dynamic simulator output files, and the data mining software developed in previous researches was adapted so as to handle the resulting very bulky temporal data bases efficiently.

Dynamic modeling hypotheses. In order to take into account the effect of uncertainty and/or errors in protection settings (delays, thresholds...) their DMs were systematically randomized in the data base generation. Similarly, the load model was also randomized in order to represent uncertainty and variability of load behavior. Moreover, in each scenario there is a random selection of some protections which are supposed to mis-operate : untimely generator tripping for over/under frequency or voltage protections, untimely line tripping for overload protections, partial non-operation of under-frequency load shedding protections, in order to represent what is happening in real life.

External disturbances. They are composed of two consecutive contingencies in the study region. They are chosen randomly (probabilities reflecting more or less their relative likelihood in real life) among the following ones :

- **Generator loss:** loss of one unit (thermal or nuclear); loss of some units of the same plant (thermal and hydro units); loss of a plant.

- **Faults on lines:** temporary fault on a line, permanent fault on a line; permanent fault on parallel circuits; two temporary or permanent faults separated by about 100ms on geographically close lines (lightning storm).

- **Faults in substations:** (bus bar fault) fault inside a substation leading to the loss of part of the substation; fault outside (but near) the substation leading to the loss of a part of the substation; loss of the whole substation after a major fault inside.

The two external disturbances are applied in sequence within a rather short time slot (less than 10 minutes) and it is supposed that there is no operator action in between. Then, the scenarios are simulated during 40 minutes after the second external disturbance in order to evaluate consequences.

Ground parameters. About 800 temporal attributes are used to describe the scenarios in the results data base. They are :

- voltage (magnitude and phase angle) of defined buses (130 attributes);
- rotor angle, velocity, and acceleration as well as excitation current and mechanical power of defined units (315 attributes);
- total active and reactive load and mean voltage in defined load areas (54 attributes);
- mean transformation ratio of "on load tap changers" in defined areas (13 attributes);
- equivalent voltage angle and frequency for the regions of the defense plan (6 attributes);
- reactive generation of the units participating in secondary voltage control (19 attributes);
- active and reactive power flows of all 380 kV lines of the EDF system, and some important 225kV lines in the study region (234 attributes);
- a listing of the discrete events happening in the system during the simulation.

Some of these attributes are to be used in order to define the severity of scenarios, i.e. measure the *consequences* in terms of loss of load and generation. The others are to be used as input parameters to criteria for characterizing the dynamic behavior of the scenarios and predict their severity as accurately and as early as possible. Those among these latter which are found to be the most informative (upon applying automatic learning to the data base) to predict the future behavior of a scenario in terms of its severity could then be used in order to monitor the system in real-time, together with appropriate decision rules extracted from the data base.

Data base generation and validation. A first preliminary data base of a few hundred scenarios was generated in early 1996, and, in August 1996 the generation of the final data base was started, using the tool described in §9.4. By the end of October 1996 the total number of scenarios simulated was about 1500, out of which about 100 were rejected.

The scenarios were simulated by Eurostag [MS92] ; this simulation program is used in dynamic security assessment studies at EDF and, with its variable integration time step, is able to simulate slow dynamic phenomena (e.g. voltage collapse) as well as faster ones (e.g. loss of synchronism). The simulations were carried out in parallel on a cluster of 12 (HP 700) workstations available at night and during the week-ends.

In order to fix ideas about the computational involvement, let us mention that in the mean a single scenario simulation required about 11 hours and produced about 200MBytes of raw output. The total amount of data extracted for the 1400 scenarios of the a posteriori data base is of 1.5GBytes, compressed.

Table 9.2 (page 192) provides a first glance at the diversity of the information in the data base.

Automatic learning. The proper exploitation of the data base is in progress. Investigations are under way in order to take the best advantage of such temporal data bases by automatic learning (some first results are given in [WLTB97a]). In particular, adaptation of decision tree induction and clustering techniques to handle temporal attributes seems very promising. Some first results are discussed in §10.4.

Notes

1. The attribute values are however normalized.

Table 9.2. Some statistics of the global DSA data base. Adapted from [WLTB97b]

Salient scenario characteristics	Min	Max	Mean	σ
CPU simulation time (s)	0	99000	38000	22000
time steps before interpolation	89	46800	3800	3071
time steps after interpolation	4	1728	145	137.2
size (MB) before interpolation	4	2140	174	140
size (kB) after interpolation and compression	32.4	3720	840	460
Number of lines lost				
380 kV	0	48	5	6.3
225 kV	0	149	9.6	18.53
Thermal units				
Number of units lost	0	15	1.15	2.1
mechanical power lost (MW)	0	13000	617	1213
mechanical power variation (EDF system, MW)	-20800	546	-1265	2445
Hydro units				
Number of units lost	0	32	2.7	5.3
mechanical power lost	0	2952	271	584
mechanical power variation (EDF system, MW)	-3039	60.5	-332	604
Load variation (MW)				
I region	-9046	654	-864	1714
L region	-8944	288	-1194	2368
EDF system	-22000	426.8	-2417	4323
Exportation variation (MW)				
EDF system to Belgium	-1527	1568	22	460
EDF system to Germany	-1832	1607	17	476
EDF system to Switzerland	-3301	830	-150	400
EDF system to Italy	-2974	1609	-48	463
EDF system to Spain	-1644	1253	-148	435
EDF system to all foreign systems	-8470	4946	-305	1626
Voltage at some buses at end of simulation (see Fig. 9.6)				
B.CARS71 (380 kV)	0	424	327	137
TAVELS71 (380 kV)	0	467	366	98
P.CORS71 (380 kV)	0	463	394	60
G.ILES71 (380 kV)	0	487	395	71
GEN.PS71 (380 kV)	0	469	395	57
VIELMS71 (380 kV)	0	449	397	45
BAYETS71 (380 kV)	0	416	387	55
VAUPAS61 (225 kV)	0	288	211	59
FLEACS61 (225 kV)	0	255	235	36

10 A SAMPLE OF REAL-LIFE APPLICATIONS

10.1 INTRODUCTION

The objective of this chapter is to provide some examples of real-life applications of the AL framework to security assessment. We aim at convincing the reader of the real usefulness of this approach by drawing our examples from various research collaborations we had with electric Utilities in order to evaluate the approach on their systems. The second objective is to provide further illustrations of the type of results which may be obtained by a sample of the automatic learning methods described in the first part of the book.

The chapter is organized as follows.

Section 10.2 briefly discusses case studies in the context of preventive transient stability, both in the context of Hydro-Québec and Electricité de France. Section 10.3 presents in some detail a very broad study in the context of voltage security assessment of Electricité de France, both preventive and emergency wise. The methodology and tools developed in this context are presently exploited by Electricité de France in order to carry out some real operational studies (see §11.1). Finally, Section 10.4 elaborates further on recent work which aims at developing a global probabilistic approach to study power systems in extremely disturbed modes. All in all, these case studies offer a representative sample of security assessment applications.

10.2 TRANSIENT STABILITY

10.2.1 A brief overview of the research carried out

Our research on the application of artificial intelligence methodologies to on-line transient stability assessment was initiated some 12 years ago. The initial objective was to assess whether and which kind of AI methodologies could be helpful to solve this problem, conventionally tackled via tedious numerical simulations, too cumbersome for on-line applications.

Since experts derive their transient stability knowledge mainly from off-line simulations, it was judged that a machine learning approach could automate this process to a certain extent. In particular, such an approach was expected to be potentially able to exploit off-line large amounts of computing powers, which were starting to become available. This motivated us to identify ID3 [Qui83] as a plausible machine learning method, able to treat large-scale problems; to assess its feasibility, we first adapted and applied it to various "academic" problems [WVRP86, WVRP87, WVRP89b, WP91].

Of course, our research was closely related to other tentative applications of pattern recognition techniques to this problem, in particular artificial neural networks. However, while the latter methods - as they were formulated - mainly relied on a small number of pragmatic features, our main goal was to stick as closely as possible to the way experts tackle the problem in real life, so as to take advantage of their collaboration and their feedback, paramount for the success of such a method. In turn, this imposed the use of standard operating parameters as attributes and required us to formulate the resulting criteria in as simple as possible manner to make their interpretation accessible to the experts. It was also deemed necessary to decompose the strongly nonlinear problem of transient stability into simpler subproblems, in order to derive reliable yet simple decision trees. Thus, we started by looking at *single-contingency* trees.

This initial research has shown the credibility of the proposed approach and consolidated the tree building methodology as it is formulated today and described in §5.3. The following step has concerned a research project started in the early 1990's in collaboration with the R&D department of EDF; the objective was to assess the feasibility of the approach in the context of the particularly large-scale EDF system. Initially, transient stability assessment was tackled for on-line purposes. But it soon became clear that this method could be interesting within the contexts of planning and operational planning as well; thus the evaluation of potentials and weaknesses and the possible improvements of the methodology concerned a rather broad field.

Within this first real-world study we were able to answer many practical questions, in particular those relating to the specification and generation of a data base, and to the improvement of the quality of decision trees to reduce non-detections of unstable situations. Later on, the research was extended to multi-contingency decision trees and considered compromises between these and single-contingency ones.

Finally, a second research project was started in 1992, in collaboration with the operation planning and control center teams of Hydro-Québec, aiming to assess the decision tree methodology in the context of their system. The long-term objective was to provide a tool for the operational planners, by allowing them to determine in a systematic way the operating guidelines for their system concerning the transient stability (and also mid-term voltage stability) limits. It was thus hoped to advantageously replace the presently used methods. While the results obtained in this research were very promising, unfortunately they did not yet give rise to an actual use of the methodology at Hydro-Québec.

Having gradually gained confidence in the methodologies of data base generation and of learning methods, we started investigating complementary features of statistical and neural network methods; they led us to make some additional tests with the data bases generated for the EDF and Hydro-Québec systems, and to develop the hybrid DT-ANN and DT-KNN techniques already mentioned.

Since in the preceding chapters we have already provided information concerning the Hydro-Québec study, below we will only describe some of the studies carried out on the EDF system.

10.2.2 EDF system

The study on the EDF system was carried out in two successive steps : (i) pilot study on three contingencies close to an important power plant; (ii) systematic multi-contingency study of the operating limits of this power plant. Below we will give only a sample of the results obtained; further details may be found in [Weh95a]. Note that the data bases generated within this context have been widely used by us and other research teams worldwide, for benchmarking various automatic learning methods.

Study system and data bases. All in all four different data bases were generated. To provide insight into the iterative "generate and test" nature of this process, we describe in detail the options concerning the two first data bases, and give some indication of the main differences of the two other ones. The reader not interested in these details may skip them, and read only the description of the base case conditions.

The results obtained within this study are described and discussed in refs. [AWP+93, WA93, WPEH94].

The considered system is an earlier version of the EDF system formerly used for operation planning studies. It encompasses the complete 400 kV grid of the French system as well as the most important part of the 225 kV network, yielding a 561-bus / 1000-line / 61-machine system. Equivalent representations were used for the surrounding European interconnections (Germany, Switzerland, Italy, Spain

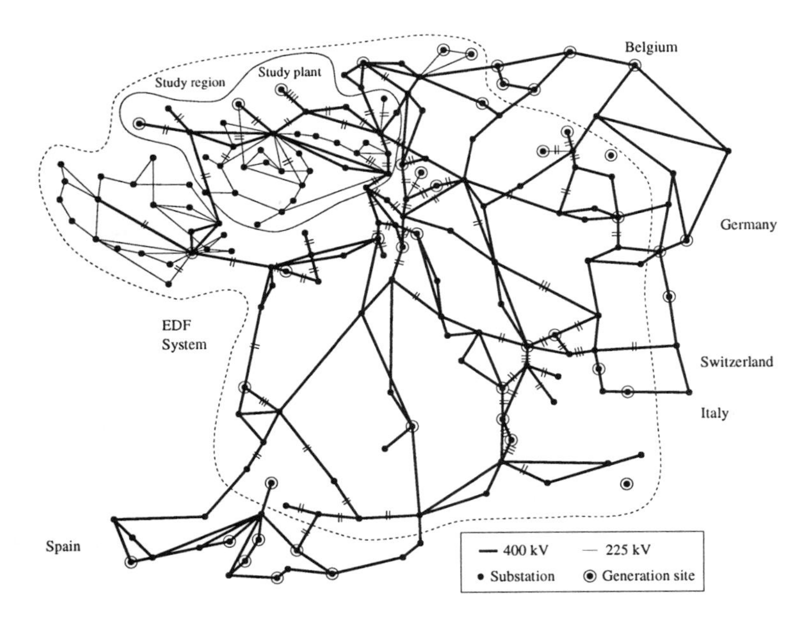

Figure 10.1. One-line diagram of the EDF system

and Belgium). The overall generation produced by the 61 (equivalent) machines corresponds to about 60,000 MW of national generation and 50,000 MW of external equivalents. Its one line diagram is sketched in Fig. 10.1.

The study concerns stability assessment of an important plant situated in Normandy (North-West part of France). This *study plant* was selected via a preliminary investigation of 60 different contingencies at the 400 kV level for 9 different operating states. Figure 10.2 describes its substation and immediate neighbors at the 400 kV level.

Data bases were generated from a base case via modifications described below. They essentially concern the "study region", but of course all load flow and stability computations were run on the entire system. This study region presumably encompasses all components liable to influence the stability of the study plant. It was determined by EDF engineers in charge of stability studies, and is composed of three large power plants (the study plant is the plant number 3) along with the surrounding substations, and about 60 lines at the 400 kV and 225 kV levels. The overall installed generation capacity of these plants is about 10,000 MW and the base case load is approximately of 5,600 MW (corresponding to winter peak load) shared among 42 different load buses.

Initial data base. The primary objective was to obtain a sufficiently rich data base, which at the same time contains plausible operating states of the region and covers

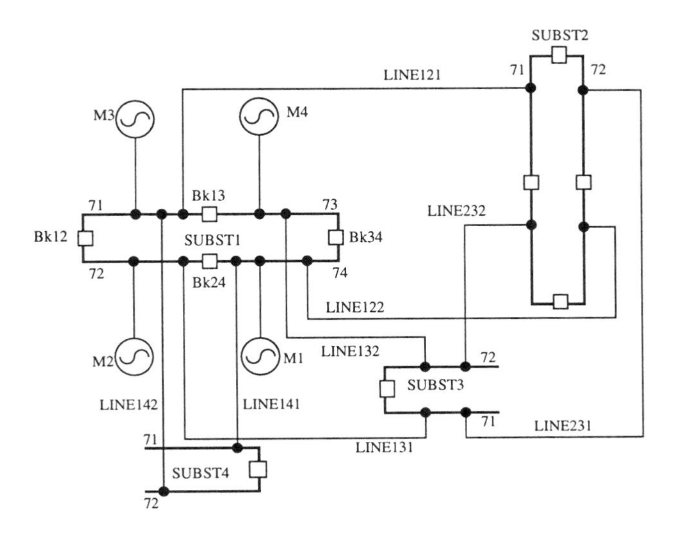

Figure 10.2. One-line diagram of the study plant substation. Adapted from [WPEH94]

as much as possible weakened situations. For this purpose, a certain number of independent variables, liable to influence the system stability were defined, concerning both topology and electrical status. For each variable a prior distribution was fixed on the basis of available statistical information about the usual situations, so that all interesting values would be sufficiently well represented in the data base. Moreover, to exclude unrealistic situations, constraints were imposed on values taken by different variables. 3000 operating states were thus randomly drawn, their stability was assessed and the values of various types of candidate attributes were computed.

For each state the following tasks were executed :

1. definition of the load level in the study region, the topology of the 400 kV regional network, local active and reactive generation scheduling;
2. building of the load flow and step-by-step data files;
3. load flow computation and feasibility check;
4. appending of the states attribute values in the corresponding attribute files;
5. computation of the CCTs of the considered disturbances via step-by-step simulations.

The following three severe contingencies have been identified, classified in a decreasing order of criticality.

Bus-bar fault : three-phase short-circuit located on the bus-bar section 71 in substation 1; cleared by opening the lines 121 and 142, tripping machine M3 and opening the breakers 12 and 13 to isolate the faulted section.

Double-line fault : 2 simultaneous three-phase short-circuits near sections 71 and 74 of substation 1, on the lines 141 and 142; cleared by opening both lines.

Single-line fault : a three-phase short-circuit on line 131 near the bus-bar section 72; cleared by opening this line.

The CCTs of the above contingencies were computed by a standard step-by-step program.

The main parameters used to draw randomized prefault operating states were topology, active generation / load, voltages, as outlined below.

Topology. It was defined by the 18 regional 400 kV lines and by the number of nodes in the 4 substations represented in Fig. 10.2.

Line outages. 10% of the states were generated with a complete (i.e. base case) topology, 50% with a single line outage, selected randomly among the 18 lines of the region; the remaining 40% corresponded to the simultaneous outage of two "interacting" lines : 40 pairs of interacting lines were defined, consisting of lines either in parallel in a same transmission corridor, or emanating from the same bus.

Study plant substation. Substation 1 was restricted to 1 node (breakers 12 and 34 were closed) if a single generation unit was in operation; otherwise it was 50% of the time configured with 1 node and 50% with 2 nodes, and so are substations 3 and 4. Substation 2 was 90% of the time configured with 2 nodes.

Load. The total regional active load level was drawn according to a Gaussian distribution of μ =3500 MW and σ =1000 MW. It was distributed on the individual load buses proportionally to the base case bus load. The reactive load at each bus was adjusted in order to keep a constant reactive vs active power ratio ($\frac{Q}{P} \approx 0.15$).

Active generation. The active power generations of the three power plants were defined independently, according to the following procedure.

1. **Unit commitment.** Given a plant, the number of its units in service obeyed a plant specific distribution. Thus, for plant 1, 0 to 4 machines may be in service, according to a priority list, and with uniform probabilities. For plant 2, the 4 following combinations were used : no unit in operation (10%); either unit 1 or unit 2 in operation (30%); both units 1 and 2 simultaneously in operation (60%). For the study plant, 10% of the cases corresponded to a single unit

in operation, 20% to 2 units, 30% to 3 units and 40% to 4 units; the units being committed were drawn randomly, under the restriction of an as uniform as possible share of the generation on the two nodes of the substation 1, if the latter was configured with 2 nodes.

2. **Active power generation.** Once again, to maximize the interesting cases the rules were plant specific. For plants 1 and 2, a random generation was drawn in the interval of the global feasibility limits of its operating units. For the plant 3 of Fig. 10.2, the first two units in service were rated at their nominal power of 1300 MW each, the next two were rated according to a random number drawn in the feasibility limits of the units. This enabled the generation of a maximum number of highly loaded situations, without losing information about intermediate, albeit less realistic cases.

Voltage profile. A simple strategy was used to produce sufficiently diverse voltage profiles, near the study plant. The EHV set-point of its operating units was drawn randomly in the range of 390 kV to 420 kV, independently of the local load level. Furthermore, the voltage set-point of plant 1 (the next nearest one) was drawn in the same range and independently. This produced a quite diverse pattern of reactive generations and flows in the study region (see below).

The above randomized modifications of the base case provide, via load flow computations, the 3000 operating states of the data base.

Incremental data base. In order to investigate the possibility of improving the decision trees by expanding a particular subtree, an incremental data base was generated for a subrange of situations corresponding to the constraints defining a particular deadend node of a tree built on the basis of the above "global" data base. This resulted in an additional set of 2000 situations corresponding to a single-node configuration of the study plant substation and the lines 132 and 141 systematically taken out of operation.

Multi-contingency data base. A third data base has been generated in order to investigate multi-contingency aspects, and, in particular, the complete set of contingencies which could possibly constrain the operating state of the power plant. Seventeen such potentially harmful contingencies were preselected by the operation planning engineers; they are detailed below. To take advantage of the experience gained with the first data base, a new set of 3000 operating states were generated on the basis of slightly different specifications. The main differences in the random sampling procedure are the following.

The study plant substation was kept in a constant single-node configuration. This simplified the stability assessment.

The line outages were restricted to the 6 outgoing lines. The probabilities were respectively of 0.1 for no outage, 0.35 for one line outage, 0.35 for two-line outages and 0.2 for three-line outages. This tended to weaken the prefault situations and hence to increase the number of unstable scenarios.

The active regional load was drawn according to a Gaussian law of μ =2500 MW and σ =800 MW. This yielded a lower level of reactive generations in the study plant and thus also weaker situations from the transient stability viewpoint.

The active generation of the units in operation in the study plant were generated according to triangular distributions instead of the uniform distributions used above. The objective was to increase the diversity of high generation situations by creating more such situations, and by distributing them on a slightly larger range of values.

Constant topology data base. Finally, a simplified data base was constructed corresponding to a constant base case topology. In this data base, the number of units in operation in the study plant was however kept variable while the total plant generation was distributed according to a triangular distribution and shared uniformly by the different units in operation. This data base was mainly exploited in benchmarking various AL methods.

Discussion. The above description illustrates the iterative "generate and test" nature of the development of an appropriate data base. All in all, the successive data bases generated for the EDF system correspond to 11,000 different prefault operating states and 18 contingencies. A total number of 65,000 CCTs have been computed and about 1,300,000 attribute values.

Incidentally, we mention that in addition to the investigations on the automatic learning methods, the data bases were extensively exploited in another parallel research project concerning the development of an improved version of the *dynamic extended equal area criterion* [EGH+92, XWB+92, XRG+93, XZG+93]. This is a typical byproduct of the data bases generated within the automatic learning framework.

Comparison of various AL methods. In parallel to our own investigations, we provided the first data base constructed for the EDF system to the researchers in charge of the StatLog project[1].

The problem considered classification of the power system into stable and unstable situations with respect to the busbar fault, which is supposed to be cleared after 155ms. There were 60 candidate attributes including regional load levels, voltage setpoints, active and reactive generation of units, power flows in important lines and logical variables describing topology. This problem was chosen as the most representative one of preventive-wise transient stability assessment. It corresponds to rather elementary observable attributes which do not play in favor of the decision tree method, which

Table 10.1. Results obtained in the Statlog project. Adapted from [MST94]

	Algorithm	*Maximum Storage*	*Time(sec)*		*Error Rate %*	
			Train	*Test*	*Train*	*Test*
	Statistical methods					
	Lin. Discrim	75	107.5	9.3	4.8	4.1
	Quadrat. Discrim	75	516.8	211.8	1.5	3.5
	Logist. Discrim	1087	336.0	43.6	3.1	2.8
	SMART	882	11421.3	3.1	1.0	1.3
	Kernel. dens.	185	6238.4	*	5.7	4.5
	KNN	129	408.5	103.4	0.0	5.2
	NaiveBay	852	54.9	12.5	8.7	8.9
	Machine learning methods					
TDIDT	Cart	232	467.9	11.8	2.2	2.2
TDIDT	Indcart	1036	349.5	335.2	0.4	1.4
TDIDT	NewID	624	131.0	0.5	0.0	1.7
TDIDT	AC2	3707	3864.0	92.0	0.0	1.9
TDIDT	BayTree	968	83.7	11.8	0.0	1.4
TDIDT	C4.5	1404	184.0	18.0	0.8	1.8
TDIDT	Cal5	103	62.1	9.8	3.7	2.6
	Castle	80	9.5	4.3	6.2	6.4
	CN2	4708	967.0	28.0	0.0	2.5
	ITrule	291	9024.1	17.9	8.0	8.1
	Neural network methods					
	Kohonen SOM	585	*	*	6.1	8.4
	Dipol92	154	95.4	13.1	3.0	2.6
	MLP bprop	148	4315.0	1.0	2.1	2.2
	Rad. Basis Fun.	NA	*	*	3.7	3.5
	LVQ	194	1704.0	50.8	1.8	6.5

obtains the best results in terms of accuracy with more sophisticated attributes. Using a larger set of candidate attributes would also have been at the disadvantage of the statistical and neural network approaches from the computational point of view.

The results obtained are summarized in Table 10.1. The first column describes the particular algorithm used; for the sake of clarity we have grouped together the methods according to the three families of algorithms discussed in Part 1. (Among the machine learning methods the first seven are of the TDIDT family : Cart, Indcart, NewID, AC2, BayTree, C4.5, Cal5.) The three following columns indicate the amount of virtual memory and of CPU time in seconds required during the learning and testing stages for each algorithm. This gives an indication of the relative performance of different algorithms, which have mostly been determined on standard workstations (e.g. SUN SPARC2). Finally, the last two columns indicate the error rates obtained in the learning and test sets. The difference between these two numbers gives an indication of the degree of overfitting of various methods.

We quote the conclusions given in ref. [MST94] :

Smart comes out top again for this data set in terms of test error rate (although it takes far longer to run than the other algorithms considered here). Logdiscr hasn't done so well on this larger data set. The machine learning algorithms Cart, Indcart, NewID, AC2, Bayes Tree and C4.5 give consistently good results. Naive Bayes is worst and along with Kohonen and ITrule give poorer results than the default rule for the test set error rate (7.4%).

We observe that the results obtained with any of the TDIDT methods are quite consistent with our own results. Indeed, the error rates range from $[1.4\ldots2.6]$ with a mean value of 1.86 %, whereas our own algorithm has obtained 1.7%. In terms of learning CPU times, the times range between $[62\ldots3864]$ seconds with a mean value of 735 seconds, which may be compared with the value of 288 seconds obtained on a SUN SPARC2 workstation with our own algorithm. On the other hand, in terms of testing CPU times our own algorithm takes about 2 seconds to complete the DT testing, which is among the fastest methods.

Multi-contingency study. Below we give a small subset of the investigations carried out in the multi-contingency study, comparing various single and multi-contingency trees. For multi-contingency security assessment the following is a sample of questions which may be raised.

What are the global stability limits of an operating condition within which it is simultaneously stable with respect to all pre-defined contingencies ?

Which are the contingencies for which a state is unstable ?

What is the overall ranking of the contingencies in terms of their severity, for a range of operating conditions ?

The first two questions may be easily tackled via single-contingency trees. However, multi-contingency trees may also be considered in order to take explicitly into account the similarities and differences among contingencies.

Within this context we may distinguish *global* and *contingency dependent* multi-contingency trees.

The former kind of trees was illustrated in Fig. 5.13 for the Hydro-Québec system. They classify an operating state as unstable as soon as it is unstable with respect to at least one contingency, without however indicating which one. Their main advantage is interpretability : they are able to provide the type of information which is necessary to an operator in order to quickly appraise the security of the system and identify potential problems and possible control actions.

The other type of multi-contingency decision trees are essentially aiming at compressing the single-contingency trees without loss of information about the identification of the dangerous contingencies.

As concerns the third question, it may be answered by various statistical analyses of the data base and in particular the so-called contingency ranking trees. The multi-contingency study results are detailed in ref. [AWP+93] ; below we merely give some examples and the main conclusions.

Simulated contingencies. We selected contingencies which are possibly constraining for the study plant. Exploiting symmetry to exclude redundant contingencies, a total of 17 faults have been defined.

12 Line faults comprising :

3 Single Line Faults (SLF) : three-phase short-circuits ($3\phi SCs$) on one of the outgoing lines which is cleared by opening the line. The fault clearing time is 90 ms.

3 Single Line Delayed reclosure Faults (SLDF) : a delayed reclosure (after 20 secs) of circuit breakers for SLFs is considered, assuming a permanent fault. The system equilibrium reached when the circuit breakers reclose is computed by running a load flow. A $3\phi SCs$ is simulated starting with this initial condition. ($\tau = 110$ms.)

3 Double Line Faults (DLF) : two simultaneous $3\phi SCs$ on the double circuit lines towards each of the three neighboring substations are considered. ($\tau = 90$ms.)

3 Double Line Delayed reclosure Faults (DLDF) : these are the DLF with delayed reclosure of breakers. $\tau = 110$ms .

5 Busbar Faults ($\tau = 155$ms) comprising :

2 Single Busbar Faults (SBF) : $3\phi SCs$ on the busbar sections cleared by isolating the busbar section, tripping the machine and opening the lines on the section. (Faults numbered 13 and 14.)

1 Double Busbar Faults (DBF) : when a busbar section is out of operation, the machines and lines on it are transferred to the opposite busbar section. A $3\phi SCs$ is assumed on this section. (Fault numbered 15.)

2 Central Busbar Faults (CBF) : when a busbar section is out of operation, if a $3\phi SCs$ were to occur on the central busbar section, up to two lines and one machine would be lost and breakers would be opened, resulting in 2 nodes at the substation. (Faults numbered 16 and 17.)

Global decision trees. The upper part of Table 10.2 gives a comparison of the test set error rates and complexities of various strategies used for simultaneous stability assessment with respect to all 17 faults. For the single-contingency DTs, the complexity is the sum of the number of nodes of all the DTs. To allow direct comparisons, the DTs built for the two strategies should be based on the same set of candidate attributes.

Table 10.2. DTs for multi-contingency stability assessment

All 17 contingencies			
No	Type of DT	P_e %	# $\mathcal{N}$
1	17 Single-contingency DTs (Set1 attributes)	6.6 %	201
2	17 Single-contingency DTs (Set2 attributes)	10.5 %	315
3	17 Single-contingency DTs (Set3 attributes)	14.7 %	214
4	1 Global DT (Set2 attributes) (see Fig. 10.3)	13.0 %	47
5	1 Global DT (Set3 attributes)	16.6 %	25

14 similar contingencies			
No	Type of DT	P_e %	# $\mathcal{N}$
6	14 Single-contingency DTs (Set1 attributes)	4.4 %	204
7	14 Single-contingency DTs (Set2 attributes)	9.5 %	264
8	1 Global DT (Set1 attributes)	5.7 %	53
9	1 Global DT (Set2 attributes)	7.4 %	41

The set1 of candidate attributes is a very rich set composed of 241 attributes including fault independent and fault dependent ones; set2 attributes corresponds to a simpler set of fault independent specific and global attributes; set3 corresponds to elementary observable attributes.

The characteristics of the tree obtained via the two strategies and for the two last lists of fault independent candidate attributes are shown in rows 2 and 4, and 3 and 5 of Table 10.2. Observe that the increase in error rate P_e of the global trees vs the corresponding single-contingency trees (e.g. 13.0% vs 10.5%) is accompanied by a dramatic decrease in complexity (e.g. 47 vs 315 nodes).

The second part of Table 10.2 shows the characteristics of the global trees obtained for a more homogeneous group of 14 contingencies, where the double and central busbar faults have been excluded.

The type of information provided by a global DT is illustrated at Fig. 10.3. It is interesting to notice that its test attributes are referring to general, fault-independent parameters of an operating state. For example, the test selected at the top node shows the influence on the stability of the number of lines in operation in the prefault phase. Other test attributes account for the total active prefault power generated or flowing through different groups of lines.

Coming back to the respective advantages of global vs single-contingency DTs, we first note that the latter often allow us to take better advantage of contingency-specific

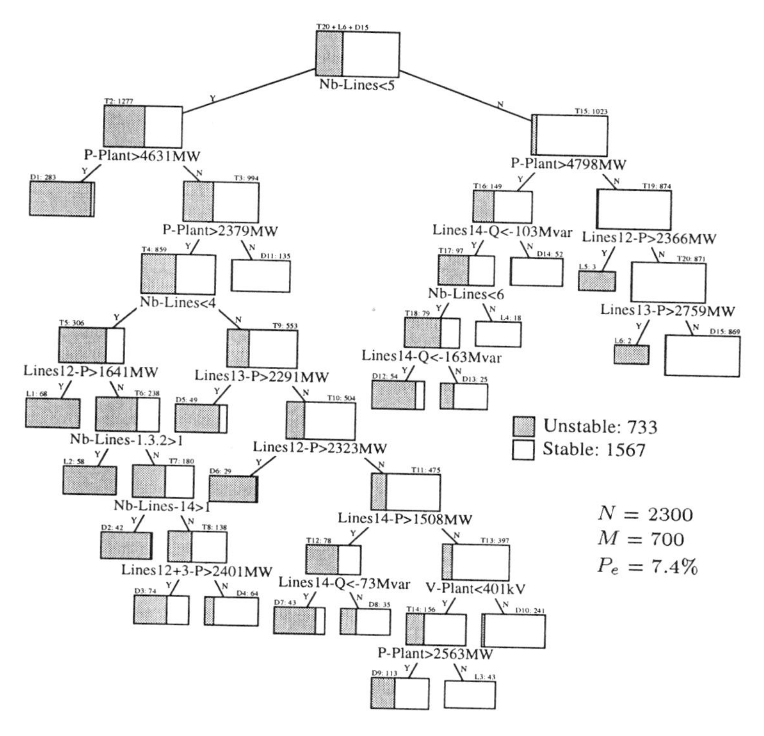

Figure 10.3. Global decision tree covering 14 contingencies. Adapted from [Weh93b]

attributes; they are able to provide richer stability information and to identify potentially dangerous contingencies. On the other hand, global trees characterize in a very simple and compact manner the structural stability limits of a subsystem of the overall power system. However, their quality depends strongly on the set of contingencies which are grouped and also on the type of attributes used. In terms of practical uses, the global trees are more likely to provide a control tool for the operator, whereas the single-contingency trees are able to express more refined information which may be usefully exploited by the engineers in the context of off-line studies and as an analysis tool for on-line operation.

Contingency dependent decision trees. With respect to global DTs, contingency-dependent multi-contingency DTs aim at telling also which contingencies are unstable under particular conditions. They therefore classify *stability scenarios* which belong to the Cartesian product set of the prefault operating states (OSs) and of the contingency set. Starting with N operating states and C contingencies this yields possibly

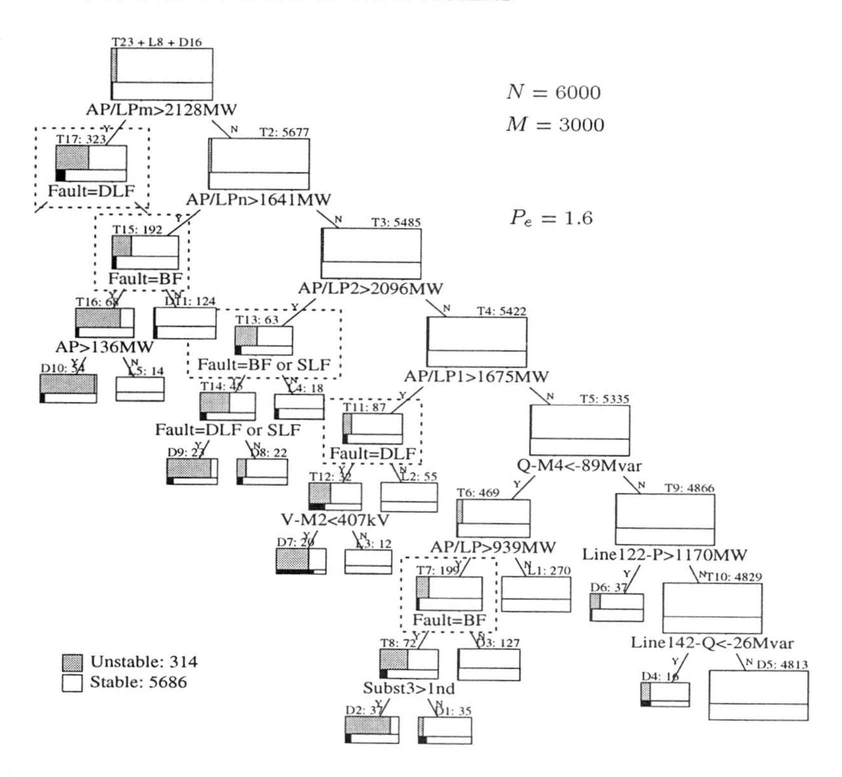

Figure 10.4. Partial view of a contingency dependent tree. Adapted from [WPEH94]

$N \times C$ stability scenarios. These are generally characterized by three types of candidate attributes : (i) contingency independent attributes characterizing the OS; (ii) OS independent attributes characterizing the contingency; (iii) combined attributes taking into account both the OS and the contingency (e.g. such as post-fault topology . . .).

One of the potential advantages of these trees is their ability to uncover and exploit similarities among contingencies. The partial tree shown in Fig. 10.4 illustrates this possibility. This tree was built for the three faults defined earlier : (i) the "busbar" fault (denoted "BF", in the tree), cleared after 155ms; (ii) the "double line" fault ("DLF"); (iii) the "single line" fault ("SLF"), both cleared after 100ms. The three contingencies together with the 3,000 operating states yield 9,000 stability scenarios : a random sample of 6,000 are used as the learning set, and the remaining 3,000 as the test set. To save space, LH and RH parts of the tree have not been represented in the figure. Note that the nodes where the retained test attribute is "Fault" are encircled by dotted line boxes.

Comparing this tree with the corresponding single-contingency ones, we observe that it has (i) a complexity of 47 nodes vs 45, the total number of nodes of the three single contingency trees; (ii) an error rate of 1.6% vs 1.7%, the mean error rate of the single contingency trees; (iii) 14 different test attributes (including the attribute "Fault") vs 18, the total number of different test attributes of the single contingency trees.

Thus, without loss of reliability, the multi-contingency tree provides a more synthetic view of the stability relationship than several single contingency trees. Moreover, similarities among contingencies are identified and highlighted by the tree (e.g. the operating states corresponding to node D10 are unstable with respect to fault BF; states corresponding to node D11 are stable for the SLF and DLF faults etc.).

Further, inspection of Fig. 10.4 suggests that, although equivalent to the information provided by a set of single-contingency trees, the information provided by the corresponding multi-contingency tree is presented in a more compact and easier to exploit fashion. This can be explained by the fact that similarities of different contingencies are exploited during the tree building so as to simplify the resulting tree. In particular, overlappings of unstable (resp. stable) regions are identified and embedded in the tree : hence, combinatorial explosion, inherent in multi-contingency control on the basis of single contingency trees, is avoided as much as possible.

Overall, we observe that a multi-contingency tree directly provides any of the following types of information :

- for a given fault (among those used to build the tree) is the considered operating state likely to be unstable or not ?
- for a given operating state, are there faults likely to create instability ?
- which conditions characterize the prefault attributes of stable operating states for a given set of possible faults ?

10.3 A DETAILED VOLTAGE SECURITY STUDY

We describe here in some details what a voltage security study carried out with this approach looks like. The case study was carried out between 1992 and 1994 on the EHV system of Electricité de France. [2]

10.3.1 *The presented study*

The problem considered is to identify the risk of mid-term voltage instabilities due to an important disturbance. In this context, the main physical phenomena which influence the system behavior in response to the contingency are (i) the load restoration thanks to either the action of On-Load-Tap-Changers (OLTC) of transformers, which act so as to restore high voltage (HV) and medium voltage (MV) to their nominal values, or actual load dynamic characteristics, and (ii) the change of reactive productions of the machines, possibly through a secondary voltage control, and their limits (over-excitation limiters action).

The presented study aims at demonstrating the approach feasibility. Its scope is thus very broad : a large diversity of network situations are considered, and both preventive and emergency aspects are treated. It is representative of an off-line security study, carried out at least one year ahead, say for the next winter. The objective could be to determine the conditions when a non-economical generation unit must be started for security purposes, or the tuning of a protection device criterion (tap changer blocking remote control...).

The power system considered is the western part of the EDF system, which has experienced a real voltage collapse in the past [HTL+90].

10.3.2 Data base specification

Network model. The power system model is a quite detailed one (see Fig. 10.5). It includes more than 1200 buses, comprising (i) the full French 400 kV network, (ii) the western part of the 225 kV level, (iii) detailed HV levels (90 and 63 kV) delimited by the dotted lines on the figure, and (iv) loads represented behind the HV/MV transformers. All EHV/HV and HV/MV transformers, which amount to 450, are equipped with OLTCs. The 320 loads connected to the MV level are sensitive to the voltage[3]. The machines of the study region all have their over-excitation limiter considered. Moreover the French Coordinated Secondary Voltage Control (CSVC) [VPLH96] algorithm is applied to the two regions covering the detailed part of the network.

Range of situations. The range of system situations is determined by the variations of two kinds of parameters.

The *primary parameters* are those suspected to have a direct, strong effect on the power system security. Let us quote for instance topological conditions, unit commitment, load level, power flows in important links or EHV voltages.

On the contrary, the *secondary parameters* are less well known and/or less controllable variables. The objective is not to interpret their direct influence on the security, but to improve robustness by making the results independent of a particular choice of values. They may be considered as noise introduced in the data base refelecting actual uncertainties. Then the automatic learning algorithms may discard the most sensitive attributes, or at least bias thresholds in order to cope with this effect. Most often, these secondary parameters concern unobservable parts of the network (low voltage levels, or external network).

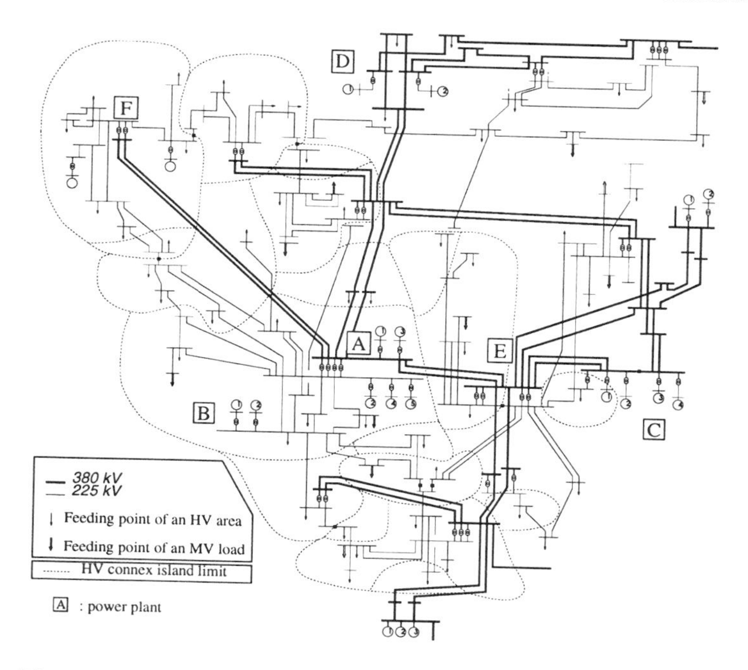

Figure 10.5. The Western part of the EDF system (study region)

Primary parameters. The topology has been treated in the same manner as other parameters. Three kinds of situations were considered, each with a given probability : full network (N), N-1 and N-2. The disconnected elements were chosen at random in predefined lists, comprising the main branches of the network shown in Fig. 10.5.

The load level has been widely varied with the aim of hitting the maximum loadability limit, to yield a sufficient number of weak situations. The variation applied homothetically to all 320 MV load buses. The base case load (a very highly loaded situation) was about 7000 MW and 2100 MVAr over this region; a uniform variation between 5000 and 9000 MW was specified.

The regional unit commitment was considered with a slight dependency on the load level : the regional importation had to stay within limits. Apart from this, the decision was taken power plant after power plant (A to D), according to fixed probabilities to have a certain number of units started. The generation dispatch was also fixed by unit type, except for the non-nuclear power plant A where a continuous distribution law has been considered, representing both cases with intermediate and nominal generation. Thus the generation is not optimally dispatched so as to ensure results validity even on unusual generation patterns, which are neither foreseeable nor avoidable when studying security problems one year or so ahead.

The voltage profile is defined via the Coordinated Secondary Voltage Control set-points. They have been varied with a Gaussian law.

Secondary parameters. In addition to these important variations, several secondary parameters were considered, and corresponding noises defined.

The load distribution within the HV network is variable. Indeed, we have considered that pure homothetical load variations might have led to over-interpretation of some particular power flows in branches feeding load regions. To avoid this, independent per-load Gaussian noises have been used in addition to the global variation.

The load sensitivities to voltage (α and β) have also been randomized. Indeed these parameters are of paramount importance when seeking an emergency control criterion. However they are very badly known; the expected issue of the noise is thus to guarantee the independence of the results from a particular choice for it. Standard uniform laws have been used for the independent, random choice of these parameters, so as to model existing uncertainty about load models.

The production dispatch among generating units in power plant A can also be seen as a secondary parameter, since this precise information was not intended to be directly used in a security rule.

10.3.3 Data base generation

The generation process.

First step : Operating point building. In addition to the random generation itself, the operating point building involves several operations, adapted to the modeling (Fig. 10.6).

1. From a base case situation, a *variant* is randomly generated according to the data base specifications.

2. Then a load-flow program is run on the base case data modified by the variant definition.

3. If this calculation converges, the voltage profile is adjusted with respect to EDF practices, using both the HV shunt compensation and the CSVC.

4. This final situation is validated : a voltage stability index is computed (the sensitivities of the total reactive power generation to a reactive power consumption, known as "reactive power dispatch coefficients").

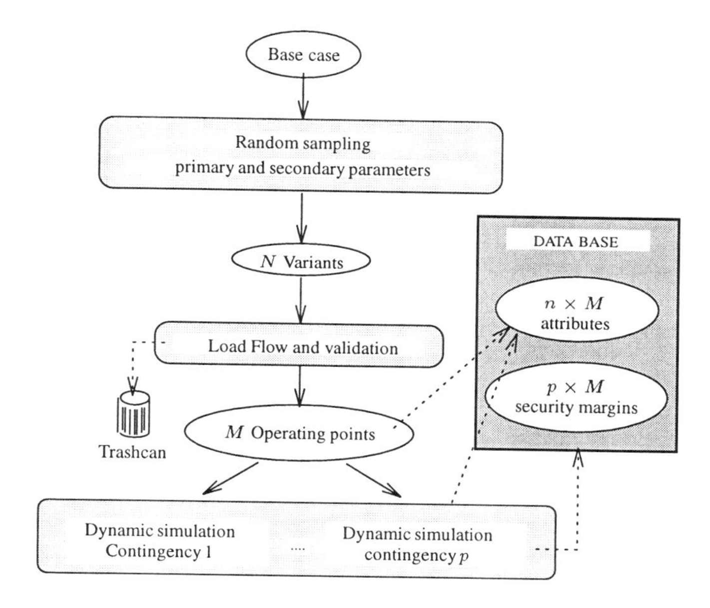

Figure 10.6. Data base generation process. Adapted from [JWVP95]

This process results in a validated *operating point*. Should any part of this process fail, then the variant would be discarded. In our study, 5,000 such operating points have been generated.

Contingencies effect and security assessment. The study aimed at determining the main problems in the region with respect to voltage security. Thus a list of the main equipment outages has been considered. 26 contingencies have been identified : losses of one or two lines, one or two generators or synchronous condensers, and busbar faults. This choice was the expert's : it has nothing to do with probabilities of any kind.

Each of these 26 contingencies has been applied to each of the 5,000 operating points, and the consecutive system behavior analyzed with the dynamic simulation tool. The post-contingency system robustness was characterized by two complementary measures :

1. the stability verdict, assessed by the use of the "reactive power dispatch coefficients";

2. the load power margin, which provides a continuous security measure expressed in terms of the additional load (P and Q) which may be supplied by the system under acceptable conditions. It is obtained by simulating a steady load increase from the post-contingency equilibrium point, while scanning the system stability;

the simulation stops when a sensitivity becomes negative, and the effective load increase is computed.

To estimate the dynamic behavior and assess the situations robustness, a dedicated simulation tool is used. The underlying idea is to simulate the interesting, mid-term dynamics (OLTCs, secondary controls, over-excitation limiters...) while filtering out the faster short-term transients (primary controls...). To this purpose, the dynamic, differential equations corresponding to the faster phenomena are simplified to algebraic ones by assuming these dynamics have reached their equilibrium point before the mid-term phenomena come into play [VJMP96]. With the limitation of being unable to highlight problems caused by the fast dynamics, this kind of approach allows drastic reduction in computing times. Nonetheless, it should be clear that the use of a simplified version of simulation tool is not a requirement of the AL approach.

A single software. The whole data base generation process has been integrated in a single software, thus rendering this operation fully automatic. One "merely" has to specify the number of operating points to generate, the random distributions for the network parameters, and the list of attributes to be collected including the stability verdicts and margins.

The result is a set of files containing the attributes and/or the verdict or margin for the range of situations. It is directly interpretable by the data mining software.

Computation time. All in all, 135,000 ($27 \times 5{,}000$) dynamic simulations have been performed, to determine the pre-contingency and the 26 post-contingency load power margins. The CPU time requirement was about 40 h per contingency, for all operating states on a Sun Sparc 10 workstation. This time is obviously indicative, since the simulation length directly depends on the states robustness : the greater the margin, the longer the simulation. Reducing this time to a per state scale yields an average of 30 s, which highlights the efficiency of the simulation tool used.

Of paramount importance are the trivial parallelization possibilities of this process : first by running several contingencies studies on different workstations at the same time, but also by allocating the dynamic simulations among as many as available workstations. The direct result is obviously the possibility to reduce the CPU time indicated above, but also to really claim the data base generation feasibility, independently of particular standard system-theory models. Should anyone want to use a more refined simulation tool, thus increasing the per-simulation CPU time requirement, then the only consequence would be an increase in the need for CPU, which becomes more and more powerful and affordable. Nearly nothing changes from the user point of view, since the process is fully automatic, and thanks to the parallel computing operating systems improvements.

A last, but not least, remark about these computation times : they have shown to represent only a small proportion of the global study length. Indeed, it takes much

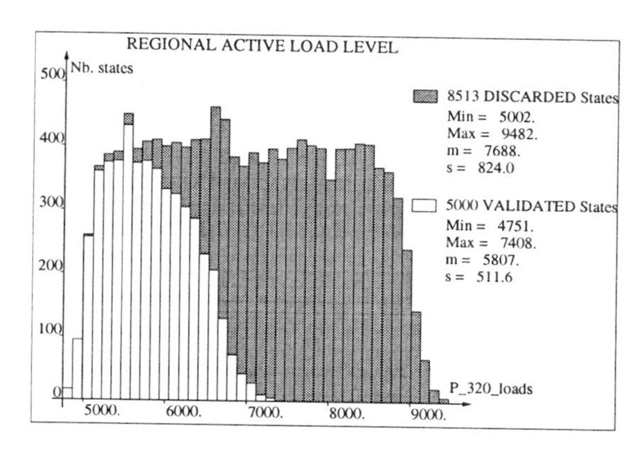

Figure 10.7. Variants filtering with respect to the load level. Adapted from [JWVP95]

more time on the one hand to think precisely of what has to be done in the study, and the way to achieve it, and on the other hand to analyze the huge amount of information contained in such a data base and get out of it the sought synthetic information. In our opinion, the engineers responsible for the study get actually much more concerned by the fundamentals of the study i.e. the physical problem, the software and data management part of it being fully automated.

Data base validation. 5,000 operating points have been validated, but 13,513 variants had to be generated for this, 8,513 having led to load-flow divergence or direct voltage stability problems. The main reason of the low percentage of load-flow convergences lies on the excessive load level specified, as illustrated on figure 10.7.

Due to correlations introduced in the random generation, other variables can suffer from side-effects. For instance, the histogram in Fig. 10.8 shows the generation distributions alteration for power plant D. In this case, the problem arises from the constraints put on the regional power importation.

But still a wide range of situations. A systematic analysis was carried out in order to assess the variability of various attributes, all in all about 400 different ones. It was found that, despite the effects of the variants filtering during the generation, the data base covers a very wide range of situations.

As an illustration, figure 10.9 sketches the distribution of the voltage in substation F, at the end of the network antenna (see Fig.10.5). One can readily see how different the situations included in the data base are : some have very low voltages (lower than 400 kV), while some others present a very high voltage profile due to a large amount of shunt compensation in operation.

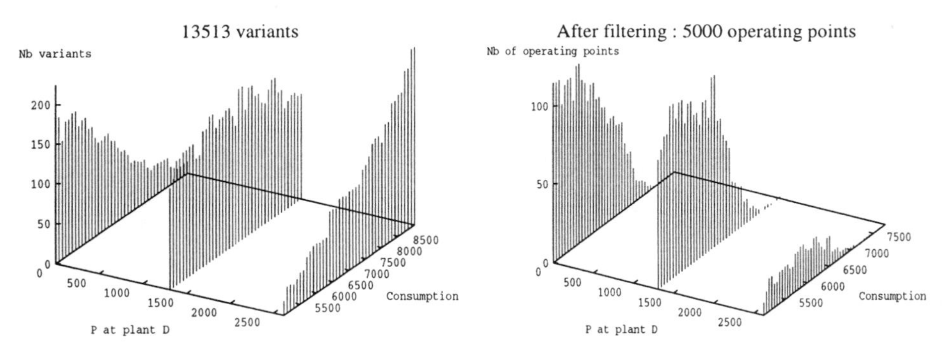

Figure 10.8. Effect of the load statistical distribution alteration on the generation at power plant D. Adapted from [JWVP95]

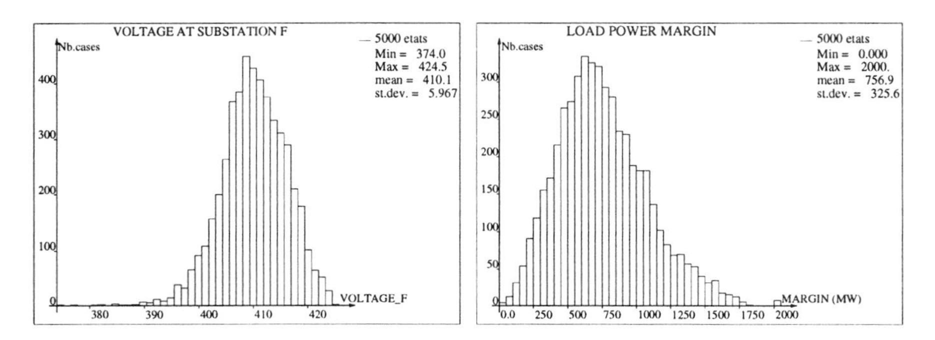

Figure 10.9. Distributions of Voltage at substation F and pre-contingency margin

This diversity appears also in terms of system robustness. The right part of figure 10.9 shows the distribution of the pre-contingency load power margin. An important feature of the data base is to have both weak cases, with small margins, and robust ones with load power margins greater than 1500 MW. In these conditions only one may hope being able to find out what makes a situation be weak.

In order to enable easy inspection of various states contained in the data base, the data mining tool was equipped with some graphical capabilities allowing to display the main characteristics of a state on the one-line diagram. Figure 10.10 illustrates the outlook of 4 different states thus visualized[4]. Furthermore, in order to be able to make some detailed analyses the data mining tool was coupled with the voltage stability simulation tool, enabling one to replay any scenario form the data base with a simple click on the mouse.

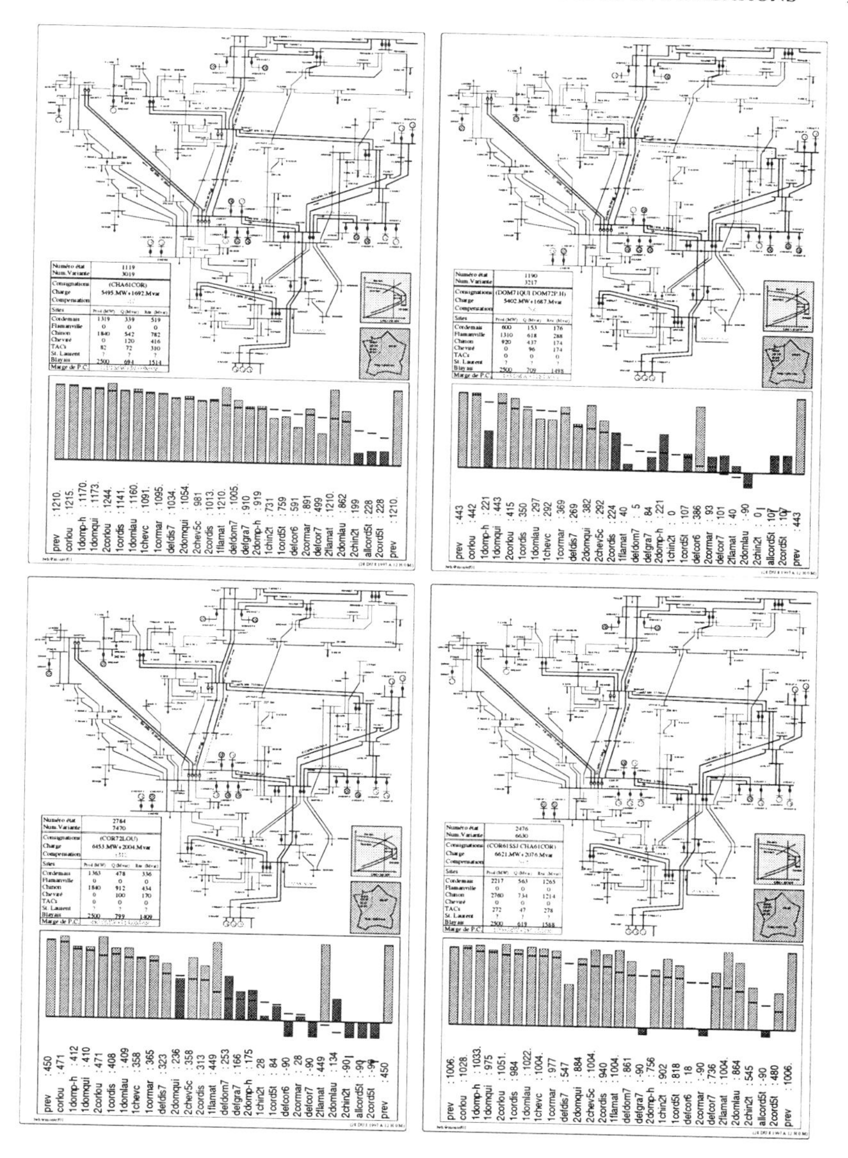

Figure 10.10. Graphical visualization of operating points

10.3.4 Data base mining

Security with respect to a contingency. Let us focus on a specific problem : determining which operating points lead to a voltage collapse when facing a given contingency, for instance the loss of one generator in power plant A.

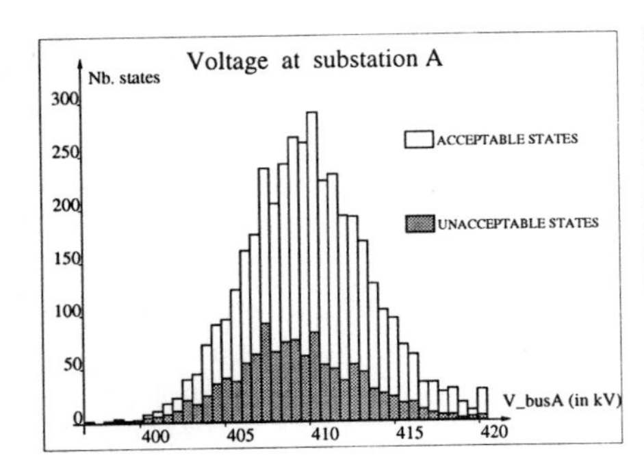

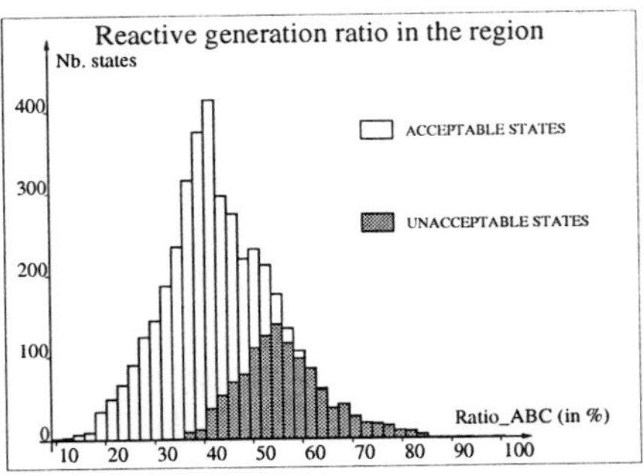

Figure 10.11. Distributions of 2 pre-contingency parameters. Adapted from [JWVP95]

To classify the states, let us declare *unacceptable* any operating point leading after the contingency to a situation with a load power margin lower than say 250 MW (including those directly unstable, i.e. without any margin).

Test defined at a glance. The first analysis merely consists of listing the system parameters liable to have an information on the operating point security : voltages, power flows, load level, generations...

A single look at their conditional distribution charts proves to be instructive. For instance, figure 10.11 sketches these histograms for the voltage at sub-station A and the reactive generation ratio (Q/Q_{max}) over power plants A, B and C. One can easily see how difficult a separation of unacceptable (grey) and acceptable (white) cases thanks to any threshold on the voltage may be. On the contrary, this task appears much easier when considering the ratio of reactive generation. Clearly, no test defined on this latter parameter will be perfect either (figure 10.12)[5].

Automatic learning by decision trees. The preceding analysis thus has shown the power of the regional reactive generation ratio as an indicator for voltage security with respect to the loss of a generating unit. Decision trees actually generalize the preceding analysis, in an automatic and multi-variable context. They can reveal complementary attributes. For instance, figure 10.13 sketches a decision tree demonstrating the complementary aspect of actual reactive reserves, and their localization among the plants.

Starting at the top of the tree, the set of situations (represented by the top box, with the proportion of unacceptable cases illustrated by the grey part of the box) is first split with respect to the reactive generation ratio, leading to one set with a great majority of unacceptable cases, and another including a majority of acceptable ones. Then the reactive reserve at plant D is used to identify a family of definitely unacceptable

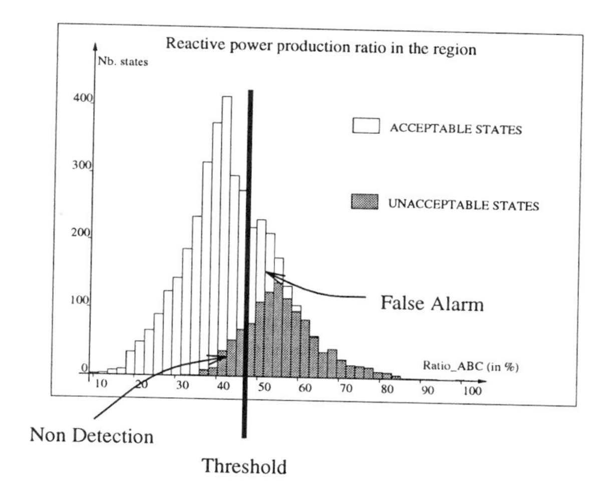

Figure 10.12. Test classification errors

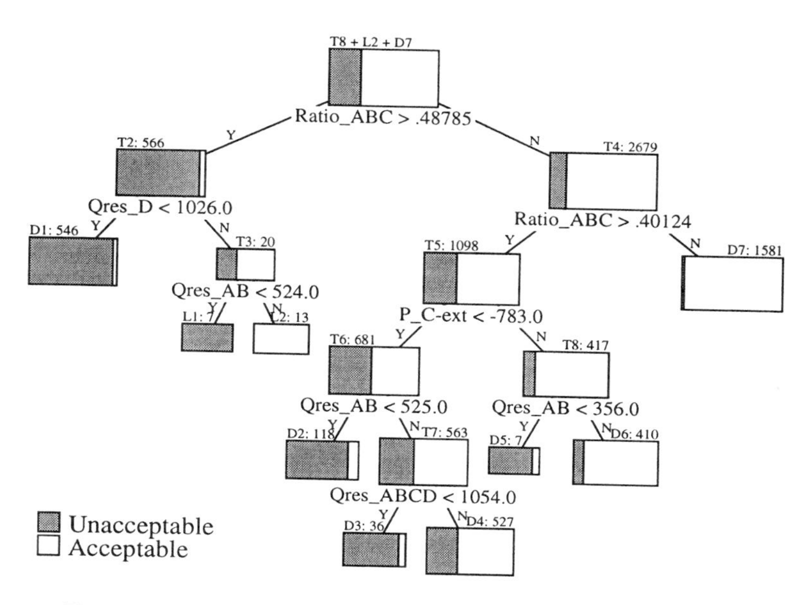

Figure 10.13. Decision tree for preventive security assessment

situations and a small set of 20 mitigated cases. The latter is correctly sorted by a test defined on the reactive reserve at plants A and B.

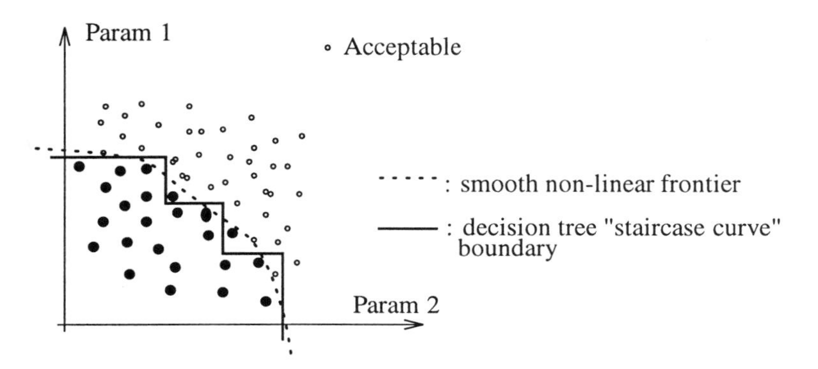

Figure 10.14. Different security boundary shapes

To use this tree as an operating rule, only the variables used in the tests have to be available. Then one traverses the tree, starting at the top-node, and applying sequentially the dichotomous tests encountered to select the appropriate successor. When a terminal node is reached, the label attached to it is retrieved.

This tree has been tested this way with 800 states which had not been used in the learning phase : the result is an error rate of 10.2 %. Will discuss this figure in some more detail below.

Test refinement by other learning techniques. Decision tree induction is thus an easy to use method, allowing to systematically explore large data bases, and extract the relevant parameters and simple rules easy to interpret.

However, the interpretability is obtained thanks to a simplification : the security boundary in the power system state space is approximated by a "staircase curve"(see figure 10.14), leading to an increase of the classification errors located nearby the security boundary.

To reduce the number of such errors, some other learning techniques can be used. Among them, the MLPs appear to be effective [WVP+94] . For our tree, the error rate is reduced from 10.2 % to 6.8 % on the same problem and even 5.3 % by using the continuous margin instead of a mere classification.

Our conclusion is thus that the artificial neural networks are relatively efficient in the task of refining the decision trees criteria, in an hybrid DT-ANN approach as described §6.3.1, but not so interesting when used on their own.

Indeed, ANNs do not explicitly select the key variables of the problem among a list which may be 200 elements long. This results in a complex, black-box result, which makes it difficult for the system security experts as well as for the operators to appraise and accept this kind of rules. Moreover, the computing time for ANN learning is in

Table 10.3. Load-power-margin computation error vs classification error

Noise standard deviation	Classification error rate	
	Gaussian distribution	Uniform distribution
10MW	1.23%	1.45%
15MW	1.71%	1.88%
20MW	2.17%	2.28%
30MW	3.33%	3.40%
100MW	10.67%	11.54%

the order of magnitudes of hours, versus a few minutes for a decision tree growing. This prevents from a systematic, extensive exploration of the data base, which however constitutes a first, necessary stage in the methodology.

So, decision tree induction is deemed to be a good central method for data base analysis and exploration, which can be completed by other learning techniques to gain accuracy, useful in the context of an on-line use. We mentioned the multilayer perceptrons, but some other hybrid approaches have also been developed to smoothen the security boundaries found by the decision trees, for instance using fuzzy sets [BW95].

An important remark about rules accuracy. The error rates given above for the security rules found may seem rather high at first glance. These figures have however no meaning by themselves.

First because the data base is not generated according to actual probabilities of system elements outages. So the 5% of cases which lead to some errors here may represent only 0.1% of the system operating conditions.

Second because an important part of the errors is merely due to numerical inaccuracies in the load power margin computations and other uncertainties. Thus, we considered a threshold of 250 MW on the load power margin as the frontier. But this margin is not 1 MW precise, far from that.

Indeed, the standard deviation of the *numerical* computation error of the margin was estimated to about 15MW. Further, for a fixed regional load level, the standard deviation of the margin variation due to uncertainty in the load *distribution* as modeled in the data base is larger than 60MW. This estimates the load-power-margin accuracy obtained via "exact" numerical computations. Using the latter to classify operating states of the data base, this uncertainty translates into classification *errors* in a way depending on the density of states in the neighborhood of the classification threshold. E.g., Table 10.3 shows the relationship between LPM computation noise and error rates, for a contingency corresponding to the loss of a generator in Plant 1, and classification with respect to a LPM threshold of 250 MW. The error rates are obtained by comparing the classification obtained by the computed margin, with the classification obtained by adding a noise term (assuming either a *Gaussian* or a *uniform* random distribution) to the computed margin. The figures are mean values from 20 passes through the data base with different random seeds.

As a conclusion, we must emphasize the fact that error rates do not have a meaning by themselves. They just provide a practical means to compare different criteria efficiencies. More important is to check the kind of errors that a security rule makes : either near the fixed boundary, and thus somewhat unavoidable, or genuine misclassification errors, which cause the rule to be inapplicable in some cases.

It is thus of paramount importance to be able to know which operating points lead to the classification errors, in order to check their actual security margin first, but also to reload them in the simulation tool in order to analyze why the rule does not correctly interpret their robustness, and have a feedback on a possible missing parameter in this rule.

Contingency severity analysis. The preceding analysis directly answers the practical question of the system security. But some better results might be expected by decomposing this problem in two successive steps : (i) evaluation of the pre-contingency state global robustness, measured by the pre-contingency load power margins; (ii) evaluation of the contigency *severity* as impact of the contingency on this robustness by measuring the difference between pre-contingency and post-contingency load power margins.

Indeed, it has been shown in our analysis that it is very difficult to find a security criterion efficient for several contingencies; and on the other hand, a given contingency may have very different effects with respect to the system state characteristics. For instance, the scatter plots of Fig. 10.15 illustrate the relation between pre and post-contingency margins for 3 different contingencies. Each point represents an operating state, and its vertical distance from the diagonal is equal to the severity of the contingency for this state. Thus the farther the location of the cloud center below the diagonal, the higher the mean severity of the contingency and the higher the spread of the points the more variable the severity from case to case.

These pieces of information about the contingency severities are summarized for the 26 contingencies by the chart of fig. 10.16, where the vertical bars show the diversity of these contingency severities. One can see how some contingencies have a rather constant effect, and how some others on the contrary lead to very diverse system robustness degradation, even for mild ones.

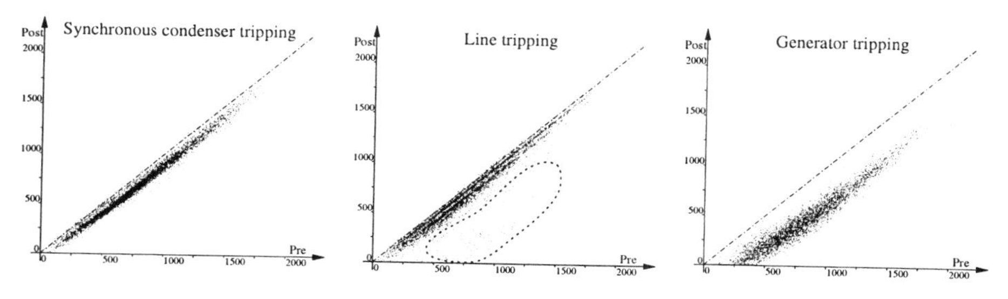

Figure 10.15. Pre- vs post-contingency security margins. Adapted from [Weh96a]

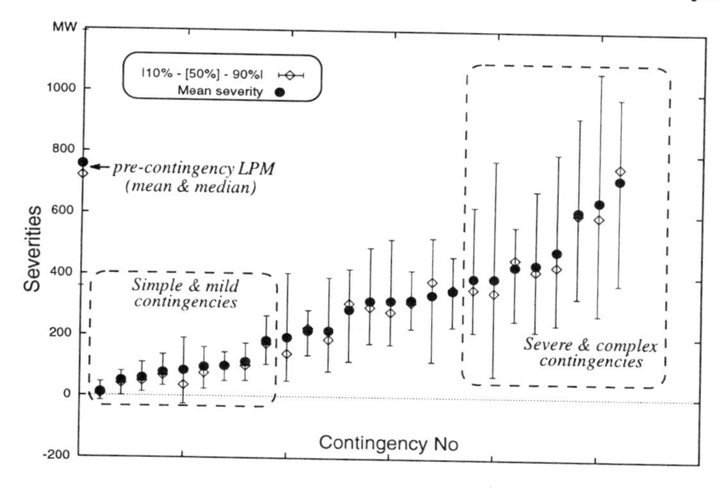

Figure 10.16. Contingency ranking by severities. Adapted from [Weh96a].

Use of regression techniques. In this context, it is interesting to be able to explain this diversity, which is expected to be linked to some very physical reasons, as an intermediate step towards the post-contingency state robustness assessment.

For this purpose, the regression techniques can be used [Weh96a]. In order to get some interpretable results, the regression tree method has been illustated on a particular contingency in §5.4.5.

For the contingency considered above, i.e. the loss of one generator in power plant A, the tree selects 15 attributes among 138 : those which are the most correlated with the severity. These comprise by decreasing order of importance : the reactive flow through 400/225 kV transformers in substation E, the total reactive HV compensation in the region, the active flow through 400/225 kV transformers in plant A substation and the reactive reserve available in this plant. The regression tree remains however quite simple, since it is composed of 18 test nodes and 19 terminal nodes.

Like for the decision trees, the model can be further refined by using the continuous modeling capabilities of multilayer perceptrons.

Using this final result to approximate the value of severity of the test states yields a mean error of -0.8 MW and a standard error deviation of 43 MW. As compared with the margin computation precision (page 219), this result is considered as very satisfactory[6].

Interest of the contingency severity analysis. The first interest of this analysis is that it is suspected to provide more physical insight into the actual security problem, since it clearly decomposes the two aspects of pre-contingency state robustness, and contingency effect on that state. It should then lead to more interpretable results, and more precise security rules.

It may also be used in complement with an on-line DSA tool providing a pre-contingency margin computation. From one margin computation, one could benefit from the regression trees to get an estimation of all post-contingency margins, and thus assess the system security.

In operation planning, this kind of rules can provide a contingency ranking, leading some more detailed studies to the most dangerous contingencies. These rules have an advantage on any other ranking procedure : they indicate which changes can cause a contingency to become much more severe.

Another kind of application : design of emergency controls.

Tap changer blocking criterion. The preceding approach can also be applied for the system protection automata tuning problem. In this case, the candidate-attributes are any kind of variable liable to be available to the automaton; the output security classification can be kept unchanged.

For instance, for a transformer tap-changer blocking device, the efficiencies of criteria defined with local voltage measurements can be compared with some others defined with remote measurements, e.g. from some plants.

The best way to achieve this task would be to consider the variables as seen from the automata, i.e temporal sets of values. In our tests however, we simplified the problem by only considering two snapshots : one before an incident, and one 25 seconds after. This delay was chosen because the tap changers begin to act 30 s after the voltages go outside the authorized dead-band. We refer the interested reader to [WVP+94] for more detailed results.

Actual design of emergency automata. The presented methodology can also be applied to investigations for an actual design and validation of various system protection schemes. Other classes of methods than automatic supervised learning may here be useful : unsupervised clustering techniques for instance allow the definition of coherent zones in the system. These zones can then be retained for load shedding schemes, or tap-changer blocking, . . .

Figure 4.7 illustrates this point by the use of a Kohonen feature map technique. The considered contingency is the loss of a generator in power plant A (This work is described in details in [WJ95]).

The feature map leads to the definition of 13 zones, according to the effect of the contingency on 39 HV voltages. From the top-left corner, we first encounter region 5 (the epicenter of the disturbance), then regions 4, 7, 3 and 2 (close to the disturbance epicenter), then two groups of peripheral regions (8, 9, 1) and (6, 10,11) and finally the remote regions 12 and 13, hardly affected by the disturbance. Thus from top-left to bottom-right corner of the map, we move from strongly to weakly disturbed load regions.

Such an automatically built map may help to find out the coherency zones in a network, and thus to design any kind of control or special protection.

In comparison to sensitivity based coherency approaches, the present technique is much more flexible and potentially much more powerful. Indeed, the sensitivity techniques are essentially providing a case by case analysis, which is determined for a given power system topology and operating state. The present approach, on the other hand, provides a systematic analysis based on a statistical sample which may be either very specific or very diverse, depending on the type of analysis sought.

10.4 GLOBAL DYNAMIC SECURITY ASSESSMENT

Let us turn back to the recent research introduced in §9.6.2 and discuss some of the first results extracted from the data base to illustrate further uses of automatic learning and data mining, in particular unsupervised learning and highlight some issues related to the treatment of temporal information.

10.4.1 Temporal attributes

In all applications discussed so far, attributes used to represent information about security scenarios as well as output information used to characterize security were scalar (either numerical or symbolic) variables characterizing the system at a given equilibrium point. This type of representation is most appropriate when studying preventive security assessment as in traditional studies. Thus, while the data base generation implies the numerical simulation of the dynamic behavior of the system as it faces various kinds of disturbances, only a very small subset of the information thus produced is stored in the preventive security assessment data bases.

However, when the design of emergency control devices is considered as well as in more general studies of the power system dynamic behavior, a great deal of valuable information would thus be lost. For example, in the study discussed in this section, to analyse the system behavior in extremely disturbed modes it was deemed necessary to store in the data base information about the dynamic behavior of the system. This is best done by using temporal attributes, which are thus functions of the scenario as well as of time. In the data base generated there are basically two kinds of such attributes : (i) temporal curves describing dynamics of voltages, power flows, angles . . . (ii) lists of events, used to record temporal sequences of discrete events, such external disturbances and triggering of various protections.

In order to exploit this information, it was necessary to enhance our data mining tool in order to be able to respresent such attributes, cope with the much larger volumes of data (about 100 times more than in classical data bases) and exploit it in data mining. At the present stage of research, however, the automatic learning methods used have not yet been enhanced. In other words, they still require the user to define scalar attributes as functions of temporal ones. However, some methods have been added to the data mining tool to facilitate this task.

We illustrate some results obtained with data mining methods using scalar attributes derived from the temporal ones, so as to give a flavor of the type of information which can be extracted and of the diversity of the dynamic scenarios stored in the data base.

10.4.2 Automatic identification of failure modes

We start our analysis by using unsupervised learning to identify different families of dynamic behavior : stable ones, and various types of breakdown scenarios. To this end we use 39 synthetic attributes describing the 1140 scenarios in terms of consequences (see also Table 9.2) : overall number of (225kV and 400kV) lines and generators tripped, amount of load reduction in various areas, variation of active power flows through interfaces with neighboring systems, voltages at various 400kV buses.

Using the K-means clustering algorithm to automatically define a certain number of classes of scenarios (see §3.6.2), we found that the data base can be partitioned into 6 main classes, described below by increasing order of severity.

1. 702 stable scenarios. They are characterized by a mean, loss of 150MW load and 6 lines; voltages close to nominal at the end of the simulation.

2. 77 local losses of synchronism. They are characterized by loss of synchronism of some hydro plants in region I (2000MW lost generation, in the mean) and tripping of some 225kV lines. In terms of load loss, the consequences remain rather limited (1200MW in the mean, mainly in region I close to hydro plants) and voltage in the 400kV system remain normal. The difference between loss of generation and load is compensated by a decrease in power exportation towards foreign systems.

3. 176 local losses of load. These scenarios correspond mainly to cascading losses of various transmission lines, and to almost no loss of generation. In the mean about

2100MW of load is lost (50% in region I, and 50% in region L) and 300MW of generation. Thus the overhead is compensated partially by primary frequency control in France (900MW) and increase in exportation (900MW). Further clustering analysis allows to identify two subclasses : 93 scenarios with loss of load in the Rhône valley and stable voltages; 83 others yielding voltage collapse and loss of load in the South-Eastern part of region L.

4. 113 region L voltage collapses. In the mean 44 EHV lines are lost, 2000 MW of generation, 6400MW of load in region L and only 1000MW in region I. At the end of simulation voltages are very low throughout region L, while they remain stable everywhere else. The large difference between loss of load and generation is compensated by primary frequency control (about 2000MW) and increase of power export (3000MW).

5. 33 region L + I voltage collapses. In the mean 17,000MW of load is lost in France (5500MW in region I, 6700MW in region L, and 4800MW outside the study region). Most of the load variation is compensated by primary frequency control in Europe. 400kV voltages drop to very low values within the whole study region : 170kV in the South-East and 230kV in the North.

6. 17 Regional loss of synchronism. Loss of synchronism extends to the whole study region and the defense plan trips lines to isolate it from the system and activates load shedding in France. Thus, very large amounts of generation are lost : in the mean 7000MW (9 thermal and 12 hydro units). As a byproduct voltage collapses in region L, where most of the load is lost (7000MW in the mean). In region I, on the other hand, voltages are low (365kV) but stable and only 2550MW of load is lost in the mean.

10.4.3 Searching for a specific kind of scenario

For example, let us search in the data base for a voltage collapse in region L, with the following specification (see Fig. 9.6) : (i) voltages in subst. A and B close to 200kV; (ii) voltage in subst. E close to 400kV: (iii) loss of load of 8,000MW evenly spread among region L and I; (iv) loss of 9 400kV and 60 225kV lines.

Using the nearest neighbor method we find that scenario No. 1439 (class 4) fits these specifications best : loss of 9,400kV and 65 225kV lines, 5 thermal units, 6 hydro units, 3,100MW of load in region I and 4,700MW in region L; increase in total exportation of 3,600MW; voltage in subst. A (resp. B, C) of 153kV (resp. 266kV, 408kV).

Let us briefly describe the chronology of events taking place in this scenario :

1. Initialization (t=0s to t=505s). At t=20s a first (bus bar) fault appears in 400kV substation B, leading to immediate tripping of 225kV lines (overload protections working improperly) and at t=20.1 the fault is cleared by permanent tripping of two 400kV lines connected to the bus bar. At t=40s 12 tap-changer blocking devices act within the 225kV subsystem close to the hydro plants. Nothing else happens until the occurrence of the second fault, at t=504s : it is a three-phase short-circuit on a critical 400kV line in the Rhône valley, leading to the loss of two major 400kV lines (a parallel line, due to overload protection misoperation, and the faulted line due to normal operation of protections).

2. Intermediate stage (t=505s to t=2000s). Given the number of circuits lost, 9 (400kV and 225kV) lines are overloaded in the study region, leading to their successive tripping between t=1650s and t=2000s. During the same period, tap changers start reacting, first in region I, then in region L.

3. Voltage collapse in region L (t=2000s to t=2904s). Upon loss of the last three lines the critical point is passed in the Eastern part of region L : tap changers continue raising their taps but load decreases. At about the same time, low voltage protections start disconnecting 56 further (mainly 225kV) lines around Lyon leading to the loss of 3100MW of load in region I. At t=2230s a second wave of 30 tap changer blocking devices are activated throughout region L, unfortunately too late. At t=2310s a further 400kV line in the North-Eastern part of region I trips on overload protection and two seconds later some generators in the Southern part trip on undervoltage protections. About ten seconds later, some other generators are lost due to overspeed protections; two further lines are tripped on overload at t=2650s and t=2900s, without notable consequences. At t=2904s the system stabilizes with very low voltages throughout region L, normal ones in region I.

10.4.4 Predicting voltage collapse

Since region L (see Fig. 9.6) is weak in terms of voltage stability, and tap changer blocking may act too late to avoid voltage collapse, it would be interesting to find an anticipative criterion to trigger emergency control. Thus, let us build a decision tree using information about machines' field currents to predict voltage collapse in the South-Eastern part of region L. Hence, we classify a scenario as a collapse if at the end of the simulation voltage in subst. A is below 304kV, and use as attributes field currents of 41 candidate machines shortly after the second fault occurrence (i.e. as early as possible after damping of transients). Note that, among the 1119 scenarios, 245 are voltage collapses (230 with voltages lower than 250kV), some local and some extending to the whole study region.

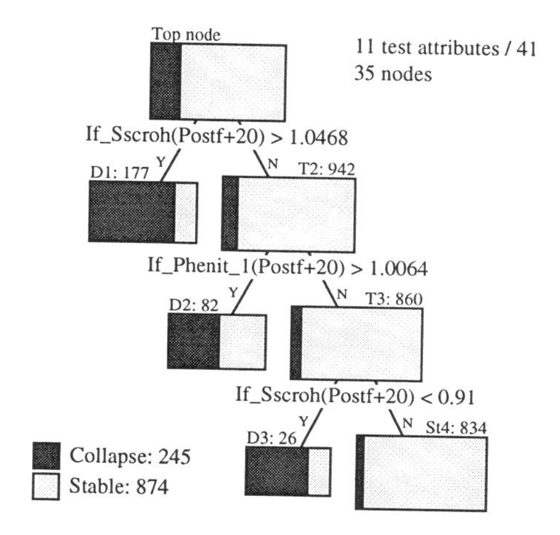

Figure 10.17. Voltage collapse prediction DT (partial view)

Figure 10.17 shows the first levels of the tree built with our method; it has 35 nodes and selects among the 41 machines the 11 most important ones. Note that the most discriminating attribute selected at the top-node (see Fig. 10.17) corresponds to a hydro unit in the South-Eastern part. Note also that, in order to avoid non-detections of collapse scenarios, the tree was biased during its building by enhancing the weight of collapse scenarios. Its reliability (estimated by cross-validation) corresponds thus to a rather low non-detection rate (6%) and a rather high false alarm rate (19%).

To see whether it would be possible to improve classification accuracy at the price of being less anticipative, we built another tree using the same candidate attributes at a later time (1000 seconds before end of simulation). The resulting tree is indeed simpler and more accurate : lower non-detection rate (2%) and reduced false alarm rate (6%). Note that it has selected field currents of 8 different machines in region L.

The preceding example illustrates the contradicting requirements between degree of anticipation and accuracy of emergency control criteria, and how data mining may be used in order to evaluate different types of compromises. Before drawing conclusions about possible criteria, many further investigations will be carried out, considering different types of attributes and different compromises. But the preliminary results already obtained look very promising.

10.5 SUMMARY

In this chapter we have screened quite a large variety of applications of automatic learning methods to a broad set of power system security problems.

In particular, we would like to stress that for these applications it is difficult to predict, at the data base specification step, which kind of methods will be the most effective, and even more, which way of applying them will lead to the most useful information : single vs multi-contingency, security class vs contigency severity, simple vs sophisticated attributes, all scenarios together or subclasses of scenarios.

In order provide enough freedom to data mining, data bases should be designed so as to contain all the information (attributes and security assessment outcomes) which might be useful later on. Nevertheless, sometimes it is necessary to enhance the contents of a data base by simulating a complementray set of scenarios, or by computing some attributes which were forgotten. The data base generation tool should be flexible enough to make this possible.

Knowledge discovery from data bases is an iterative process, leading to successive trials and assessments of results. We believe that it requires both problem specific expertise and a good understanding of the capabilities of the various automatic learning tools; and a good deal of imagination is quite helpful, also.

Notes

1. StatLog is a research project funded in October 1990 under ESPRIT program of the European Community, and concluded in June 1993. It involves 6 academic and 6 industrial laboratories which compared about 20 different automatic learning methods on about 20 different real-life datasets. The results of this project are published in [MST94]

2. The material presented in §10.3 is updated from orginal material provided by Yannick Jacquemart from Electricité de France. We take full responsibility of the possible mistatements introduced.

3. $P = P^0 \left(\frac{V}{V^0}\right)^\alpha, Q = Q^0 \left(\frac{V}{V^0}\right)^\beta$

4. Note that the bar diagrams in the lower part of each graphic of Fig.10.10 represent the severity of the 26 contingencies simulated for each case.

5. Obviously, some safety margins may be considered, especially to reduce the non-detection rate, but this results in a false alarm rate increase.

6. Regression trees for the two other contingencies shown at fig. 10.15 lead also to very satisfactory results, and can be found in [Weh96a].

11 ADDED VALUE OF AUTOMATIC LEARNING

As we have seen, the application of automatic learning to security assessment may appear to be rather involved, and its use in real life will imply rethinking security assessment practices and learning to master new tools and techniques. There is certainly no point in applying such a methodology to problems which are already solved or are in the process of being solved in the near future. Clearly, many engineers in Utilities will be skeptical and reluctant in using any new technology without strong motivations, especially as technical concerns tend to become secondary with respect to more commercial ones.

This chapter aims at clarifying the contributions of automatic learning to power system security assessment, in terms of its complementarity with existing analytical tools. It contains two brief sections.

Section 11.1 reports on the motivations and conclusions drawn from a real operational study recently carried out by EDF engineers. This gives the viewpoint of an end-user in the context of a very specific application.

Section 11.2 on the other hand discusses on a more general - long term - perspective the potentialities and motivations for using the methodogy, from the viewpoint of the author of this book.

11.1 AN OPERATIONAL STUDY CARRIED OUT ON THE EDF SYSTEM

The study was initiated by the engineers of a regional control center in order to define operating rules which are used for operating the system in a secure fashion.[1] The rules which were used until then had become obsolete, and needed to be refreshed in order to take into account the changes which had taken place in the system since they were designed.

Thus, a study similar to the one described in §10.3 was launched in summer of 1996, in order to determine the new operating rules. However, while the study described in §10.3 had the broad objective of defining a methodology for voltage security assessment and exploring various features, the present study had a very precise objective of defining operating rules which can be used by operators to make decisions in real-time. Furthermore, an important requirement was that these criteria should be as robust as possible, yet simple enough to be of practical use.

The effective time required to carry out the study was of about 5 person.months, including the data base specification, generation, data mining and report writing. The data base was composed of 10,000 different operating states yielding more than 100,000 dynamic simulations, which were carried out within about 15 CPU days. Most of the time was actually devoted to the data base specification. Hence, it is judged that if a similar study would be carried out a second time, e.g. in order to take into account the effect of a new equipment, the data base specification step would be significanlty shortened and the study would not take more than 2 person.months, overall.

Concerning the rules which were extracted by decision tree induction, they were found to be very simple and at the same time quite accurate. In particular, the method allowed to identify attributes different from those which were used before (in the "manually" designed rules) and significantly more effective in terms of predictive accuracy. In short, the results were deemed very positive. Notice also that during the data base building various uncertainties were taken into account, by randomizing parameters concerning load models, field current limits and voltage set-points of secondary voltage control. This allowed to increase the robustness of the resulting guidelines with respect to these uncertain parameters.

One of the main advantages reported by the engineer in charge of the study, was that the decision tree approach allowed him to identify the most discriminating attributes in a rather direct and objective way, while in traditional studies this would be based only on intuition. Note that the resulting rules were compared with the previously used ones, and found to be more reliable and applicable to a much broader range of operating conditions. As a conclusion, the application of the automatic learning framework in this real study has allowed to derive better security rules, which are robust with respect to various uncontrollable or uncertain parameters, and which are systematically validated.

In terms of complementarity with respect to numerical simulation we can make two other comments. First, for the time being there is no on-line dynamic simulation tool available in the regional control center concerned. Thus, in the present situation automatic learning allows to bridge the gap between the control room and tools available in the off-line study environment. Second, while, sooner or later, we may expect that such tools would become available in the control room, the information provided by the rules derived by automatic learning will remain complementary. Indeed, they provide direct information on the possible preventive control actions which would enable the operator to bring the system back to a secure operating state, whenever deemed necessary. This would be much less straightforward with the sole numerical simulation tools.

11.2 SUMMARY

We sum up by first restating our objective in this part of the book.

In the first part of the book we considered a *tool-box of automatic learning approaches*, starting with an intuitive description of complementary methods in the context of dynamic security assessment, and further discussing the main technical questions which underly their proper use.

In the second part, we concentrated on *practical application concerns* to power system dynamic security assessment. Thus we have screened the diversity of such problems and practical DSA environments in order to highlight possible uses. We have also discussed in detail the technical aspects of building security information data bases, further illustrated by practical case studies. These paramount but very time consuming aspects are generally hidden in the literature on the subject.

Finally, in order to make the preceding "theory" become credible, Chapter 10 has shed some light on the way these methods can be really applied in practice.

Maybe it can now be understood why the application of automatic learning to DSA, which was envisioned almost thirty years ago, starts only today being applied in real life. We think that there are several "complementary" reasons to this state of affairs.

Of course, as we already mentioned, technology was not ripe thirty years ago.

Then, ten years ago, when technology became mature enough, there was no methodology, and it took a few additional years of research to come up with what he have presented.

The last, maybe more difficult requirement is to convince Utilities of the usefulness of this apparently very bulky framework. Indeed, to adopt the methodology needs to change the way of thinking and of carrying out security studies, and this is possible only with a strong enough motivation. Even in the case of Electricité de France, we believe that it will take some more years before the methodology will be used in a fully systematic way in all the places where it has already shown to be useful, not to say in other contexts.

Thus, we hope that to those who have reached this part of the book we have been able to show at least some of the practical possibilities of this very powerful framework. The final decision whether to take advantage of the automatic learning framework to DSA lies in the hands of the Utility engineers. We believe that the rapidly changing economical and technological contexts of today will probably encourage some far seeing ones to start considering this framework as an actual alternative to present day practice.

Let us come back to the main complementarities we claimed in chapter 7.

11.2.1 Computational efficiency, for real-time applications ?

This was the original motivation in the late sixties. With todays computers, and with the expected growth in the near future, it becomes more and more common to hear that on-line DSA, based on full fledged numerical simulation, is now feasible. So why should be bother with automatic learning ?

There are several reasons. First, computational efficiency does not only concern CPU times, but also data requirements. From this point of view, we have shown that automatic learning can help us to make the best use of available information, in a much more direct and efficient way than we could do with only numerical techniques. Second, power systems tend to rely more and more on automatic special stability and emergency control systems. In this context, the time available for decision making is further reduced by a factor of hundred of more, and reduced data requirements are paramount. We have also illustrated how automatic learning can be used to exploit numerical simulation tools in order to design such automatic devices.

11.2.2 Interpretability and physical insight ?

Automatic learning allows one to carry out ranging studies in a systematic way, where the effect of several parameters on system security is studied at the same time. With today's available computing power, traditional, manual approaches become ineffective, because the engineer cannot analyze systematically the effect of more than two or three parameters on a system, let alone take into account modeling uncertainties.

The possibility to carry out broader ranging studies will allow the engineers to have a more *anticipative* view on potential problems. This is very different from the present day practice, which generally handles new problems *after the fact*, namely after that some undesirable consequences have already been observed on the system.

11.2.3 Management of uncertainties ?

Even if models are improved on a daily basis, and even though it becomes possible to simulate 10,000 bus systems with acceptable response times, nobody would reasonably claim that numerical simulations always provide the right answer. Looking at past experience, coming from detailed analyses of large blackouts, it is striking that among the "causes" of these latter one finds always that something didn't work as expected, be it a relay which mis-operated, a tree (a real one, not a decision tree!) which has grown too fast, an operator who did not behave as expected, a set-point which was different from prescriptions, loads which were not modeled appropriately...

While the situation will (hopefully) improve, uncertainty concerning the actual physical behavior of such complex systems is really unavoidable. Especially, as the present tendency is towards more independent human decision making, systems will become more uncertain. The only way to cope with this problem, is to make more simulations by relaxing assumptions on the dynamic models used (i.e. by randomizing their uncertain part). We have provided several illustrations of how this may be carried out in a systematic way within the automatic learning framework.

Finally, anticipating on the next chapter, another very strong motivation for using automatic learning is the need for probabilistic security assessment. While, up to now security assessment is mainly carried out in a deterministic way, it is felt that probabilistic security assessment will be necessary in order to reduce operating costs while preserving reliability, so as to survive the new competitive market structures of the power industry.

Notes

1. The information reproduced here is based on a private communication with the engineer in charge of the study. Further results will be published in the near future.

12 FUTURE ORIENTATIONS

In this chapter we discuss the main topics for further research and development in order to improve the scope and enhance the features of the approach. We start by describing probabilistic security assessment and its connection with the automatic learning framework. Then we discuss the need for enhancing automatic learning methods in order to exploit more efficiently temporal data. We conclude by proposing a software architecture which should be developed in order to enable the systematic use of the approach.

12.1 TOWARDS PROBABILISTIC SECURITY ASSESSMENT

The probabilistic nature of power system security (dynamic and static) has been well recognized since the early days of modern power system operation and control. However, in spite of numerous attempts since the early eighties (see e.g. [BK80, AB83, WTY88, VPGN94, MFV+97]), probabilistic security assessment has not yet seen actual applications. Nevertheless, in the last few years some electric Utilities have started considering it as a potential alternative or complement to existing deterministic practices [LWCC97, UPK+97] .

Our presentation of probabilistic security assessment in §§12.1.1 and 12.1.2 is based on Chapter 4 of the final report of CIGRE TF 38.03.12 on "Security assessment". The original text of this chapter was written by the author of this book, but the contribution of all the other members of CIGRE TF 38.03.12 was instrumental in clarifying many ideas. Below we focus only on those aspects which are relevant to our discussion of automatic learning based security assessment. The interested reader may refer to the full report which will be published in ELECTRA during 1998.

12.1.1 What is probabilistic security assessment ?

The aim of probabilistic security assessment (PSA) is essentially to make systematic use of the information available in any decision making context. Indeed, much of the information available to the decision maker is of a probabilistic nature, such as future load patterns, next contingencies, neighboring system states, long term economic and social contexts to quote only the most straightforward ones. Furthermore, among the aspects usually considered as deterministic, most suffer from various sources of uncertainty, resulting for instance from limitations in modeling and measurement systems, and hence are better modeled by (subjective) probability distributions than deterministic default values.

The probabilistic approach may be described as a sequence of steps leading to the evaluation of alternative decisions, in terms of their impact on the operating (and investment) costs and economic risks generated by disturbances. Notice that the tradeoff between normal operating (and investment) costs and security will be a consequence of the choice of a severity function, which is further discussed below.

A main outcome of this formulation is that the impact of a disturbance on decision making is now weighted by its probability of occurrence and its impact in terms of global economic consequences, measured by the severity function. The probability of occurrence may vary with equipment location and configuration as well as with meteorological conditions. The economic consequences may be small (e.g. loss of a small plant without important variations in frequency and voltages) or large (e.g. a partial blackout). This is in strong contrast with the classical deterministic procedures, where all disturbances are considered on an equal weight basis, and severity assessment provides typically a Yes/No type of information.

Table 12.1 describes a theoretical framework for probabilistic security assessment. Various security scenarios are defined by the conjunction of three components: a pre-contingency equilibrium state (OP), a modeling hypothesis (MH) defining structure and parameters of static and dynamic models, and a sequence of external disturbances (ED) which are supposed to initiate the dynamics. Given the precise definition of a security scenario one can in principle determine its severity (or conditional risk), which evaluates technical and/or economic consequences of the scenario. Since in all practical decision making environments at least some of the three scenario components

Table 12.1. Theoretical probabilistic security assessment framework

1. Assumption of a prior probability distribution of static and dynamic models (MH) of the power system and of its possible pre-contingency equilibrium states (OP), depending on the decision making context (ctxt) and the information (info) available therein:

$$p(\mathrm{MH}, \mathrm{OP}|\mathrm{ctxt}, \mathrm{info}).$$

2. Assumption of a conditional probability distribution of all possible disturbances (ED) according to the context and the information at hand (this could model faults as well as random load variations):

$$p(\mathrm{ED}|\mathrm{ctxt}, \mathrm{info}, \mathrm{MH}, \mathrm{OP}).$$

3. Definition of a severity function, which evaluates the severity of a particular scenario in terms of its consequences (e.g. expected economical consequences taking into account the various costs of load-shedding, voltage and frequency degradations, brown-outs...):

$$\mathrm{severity}(\mathrm{ctxt}, \mathrm{info}, \mathrm{MH}, \mathrm{OP}, \mathrm{ED}).$$

4. Evaluation of the overall risk as the expectation of severity $\mathrm{Risk}(\mathrm{ctxt}, \mathrm{info}) = \int\int \mathrm{severity}(\mathrm{ctxt}, \mathrm{info}, \mathrm{MH}, \mathrm{OP}, \mathrm{ED}) * p(\mathrm{ED}|\mathrm{ctxt}, \mathrm{info}, \mathrm{MH}, \mathrm{OP})] * p(\mathrm{MH}, \mathrm{OP}|\mathrm{ctxt}, \mathrm{info}).$

5. Evaluation of a decisional alternative by summing its corresponding Risk on the one hand, and its operating (and investment) costs on the other. (Note that in practice this latter are also of probabilistic nature and the value used is actually an expected value.)

are uncertain, decisions should be based on the expected severity (i.e. the risk) by taking into account (at least in theory) all possible scenarios and weighting their severity by their probability of occurrence.

Notice that the probability distributions of power system states, models and disturbances depend on two conditions, which we have distinguished on purpose: context (ctxt), to distinguish between planning, operation or management environments; and information (info) to account for the dependence of decisions on the quality of prior knowledge about models, forecasts, disturbances, etc.

Thus the probabilistic approach allows - in fact obliges - the user to model his degree of ignorance on the various aspects. In particular, not only the distribution of power system states and dynamic models of his own power system, but also the level of accuracy of knowledge about states and behavior of the environment (neighboring Utilities, customers, etc.) can in principle be taken into account.

Finally, a very appealing characteristic of the probabilistic approach is that it is completely general and may in principle be used in an integrated fashion at the various steps of decision making. From long term system planning to real-time decision making, the same methodology can be used, only the shape of probability distributions and possibly the definition of severity functions would change.

The application of probabilistic analysis to security assessment attempts to evaluate the expected overall risk level associated with each decisional alternative which can be made in order to ensure system security so as to select the most appropriate one.

Thus, both the severity of a scenario which may happen and the probability that this scenario happens are taken into account in the definition of the risk.

As noted above, the probability distribution of pre-contingency states and models p(MH,OP|ctxt,info) as well as disturbance probabilities p(ED|ctxt,info,MH,OP) will depend on context (ctxt) and on the information (info) available at the time of decision making.

The severity itself is a function of the scenario, but it depends also on the decision making context and the information available. Notice that, for the sake of simplicity, we have assumed above that this is a deterministic quantity. However, in practical situations outage costs will depend on many uncertain factors, like for example the time taken for load restoration as well as extraneous factors influencing the customer perception of service interruptions.

Probability distributions and severity functions, hence the risk, depend of course on the decision alternative which is considered.

12.1.2 Modeling the available information

In the probabilistic approach, the modeling steps 1 and 2 in Table 12.1 explicitly take into account information accuracy. In particular, the information which is not perfectly known in a given context may be randomized appropriately (e.g. unobservable parts such as neighboring Utilities or badly defined modeling aspects such as load behavior etc. should be represented by "uninformative" probability distributions), in order to yield robust decisions.

On the other hand, by using sharper probability distributions corresponding to more precise data, it is possible to assess the value (in terms of impact on the decisions) of better information, and thereby allows one to justify economically the gathering of new data, for example by improving measurement systems, by exchanging information with other Utilities, or by refining models and improving model parameter estimates.

Thus, the amount of information available about the system clearly influences the expected risk. It is therefore important to be able to study the influence of the quality of information available in a given context on the quality of the decisions which will be taken therein.

Notice that usual practice in deterministic security assessment consists of replacing unavailable information by default values: conservative values and hypotheses for those parameters that are expected to have a significant influence, expected values otherwise. In the probabilistic approach, it is possible to use instead a properly chosen probability distribution, thereby avoiding biased evaluations, and, when new data become available the probability distribution only need to be updated.

For example, in on-line security assessment if the operator becomes aware of the fact that a lightning storm is crossing an important transmission corridor of his system, he will significantly update his expectations of disturbance probabilities. Such knowledge allows him to justify economically some decisions in order to reduce the risk of the system with respect to contingencies in this corridor.

Similarly, in a power system with a large amount of independent producers and dispersed generation it is well known that the lack of information available about these system components may lead to difficulties and increase costs, since robust decisions must be taken with respect to the unknown behavior of the uncontrollable part of the system. Again, the probabilistic approach would enable one to assess the economic impact of various alternatives. Similarly, the effect of exchanging (or hiding) system data between different Utilities can also be modeled in this way.

12.1.3 Bridging the gap with the automatic learning framework

Due to the nonlinear nature of power systems, severity (whatever its precise definition) is essentially a nonlinear, possibly discontinuous function of the various parameters defining security scenarios. Therefore, analytical methods would probably be of a very limited scope in the risk computation and we believe that Monte-Carlo simulation techniques will be required. In spite of the very bulky nature of these simulations we believe that further growth of computing power and development of appropriate software will make this possible.

Monte-Carlo simulation, in its most elementary form, would imply random sampling of security scenarios according to the prior probability distributions assumed for OPs, MHs and EDs, together with the simulation of the power system dynamic behavior for different decision alternatives, in order to assess the corresponding severity, and to compute the sought expected values and standard deviations [Rub81].

However, given the fact that scenarios of high severity will be extremely rare events, it would be necessary to run huge numbers of simulations before obtaining reasonable estimates of the risk. Thus, we believe that this approach may become feasible only if appropriate variance reduction techniques are applied, in particular techniques able to sample more effectively the scenarios with higher severity.

Actually, although we did not call it variance reduction, this is basically what we do in the context of data base generation in the automatic learning framework : probabilities of random sampling of those parameters which are expected to influence security significantly are biased in order to increase the proportion of insecure scenarios. Thus the tools developed for data base generation can be exploited for Monte-Carlo based risk assessment, provided that the ratio between actual probabilities of scenarios (corresponding to the prior probabilities assumed in the experiment) and the biased ones (corresponding to the random sampling laws used to generate the data base) are recorded in the data base.

Thus, we may conclude that the tools used today for the application of automatic learning to security assessment will also be useful in future probabilistic security assessment [Weh97b].

Furthermore, the application of automatic learning itself may provide interesting information about the parameters most strongly influencing the severity variance (e.g. a regression tree could identify key parameters, easy to compute, which explain most of the severity variance). Such an approximate information may be used in turn in order to reduce variance in the Monte-Carlo approach. For example, it may be used in the following scheme

- generate a first data base to build an approximate model of severity by automatic learning;
- generate a very large data base and use the approximate model (easy to apply) to compute its expected severity approximation, with good accuracy;
- generate another, smaller, data base to estimate the expected approximation error.

The first step can be carried out in a preliminary off-line study. The second step does not require any numerical simulation to determine the severity. It requires only the computation of the attributes used in the approximate model derived in the first step, which may be chosen to be easily computable. The last step will in principle require a much smaller number of simulations, since, if the approximate model is good enough, approximation errors will be small, hence also their variance.

This strategy conforms to the general scheme proposed in [PMOP92, Shr66], where however the analytical approximation is replaced by a synthetic model derived by automatic learning.

Notice also that automatic learning is the tool required to extract from the Monte-Carlo simulations criteria which can be used in the on-line environment for decision making. Such criteria would use as attributes those parameters (among those used to sample the simulation scenarios) about which information will become available in the on-line environment. Thus, for example, regression trees could be built off-line so as to relate expected scenario severity to operating point characteristics and disturbances. Then, when information about these parameters becomes available in the control room (from SCADA and meteorological information systems) the regression tree might be exploited to estimate the conditional risk. Some of the operating point parameters would correspond to possible preventive control actions, some to different emergency control actions. Some other parameters would be uncontrollable but observable, such as the load level, available equipment, and disturbance probabilities. They would be injected into the tree in order to find out how to modify the controllable parameters so as to yield an optimal compromise between operating cost and risk.

12.1.4 Incorporating DSA criteria in probabilistic planning studies

Another possibility to exploit the results of automatic learning in probabilistic studies consists of using the criteria derived by automatic learning as constraints in the probabilistic planning tools used in some Utilities today.

These tools allow one to estimate expected operating costs and load curtailment during a certain future period. At the present stage, they are not able to take into account dynamic security limits [DM86]. However, it would indeed be possible to exploit dynamic security limits derived by automatic learning, and this would allow one to determine expected costs and load curtailment quantities in a more realistic way.

12.1.5 Conclusion

The automatic learning framework to security assessment, even though it is presently used mainly for deriving deterministic criteria, is essentially providing the tools and methods which will be necessary for future probabilistic security assessment.

What is presently missing in this framework is a methodology for defining actual probability distributions and for taking explicitly into account the biased sampling of security scenarios in the derivation of output information. Further work is required on the one hand to gather statistical information about various parameters and exploit this in the data base generation process in a transparent way. In addition to this, much further work is required in order to estimate outage costs from dynamic simulations.

12.2 HANDLING TEMPORAL DATA IN AUTOMATIC LEARNING

Our recent researches (see §10.4) suggest that in many of the most interesting applications of automatic learning (e.g. emergency control system design), security information data bases will contain a great deal of temporal attributes.

However, while existing automatic learning methods are good at exploiting non temporal (scalar) attributes, they are clumsy with temporal data.

Adapting a method such as decision tree induction to the treatment of temporal data will involve developing families of candidate tests which operate directly on temporal curves. An example of such a possibility would consist of searching for temporized thresholds, which would amount to defining a threshold on a given attribute together with a temporization. This could be achieved by a two-dimensional optimization technique, similar to the one used for determining linear combinations (see Table 5.9).

Furthermore, when developing a decision rule for emergency control there are two contradicting requirements. On the one hand, it is certainly important to avoid false alarms. At the same time, it is important to make sure that the rule detects instabilities as early as possible so as to enable timely triggering of mitigation schemes. Indeed, constructing a decision rule which would be able to "detect" instability after a voltage collapse or loss of synchronism has already occurred, would be at the same time easy and quite useless. Thus, during automatic learning, when comparing alternative decision criteria, it will be necessary to take into account both their degree of anticipation and their discrimination capability, so as to reach a good compromise. This will require the development of different quality measures than those used in the classical applications with scalar attributes, and is a topic for further research.

In addition, assuming that the above fundamental questions are solved, it will be probably necessary to work on improving computational efficiency. Indeed, temporal data bases are typically two orders of magnitude larger than non temporal ones, and computational problems may become intractable without parallel computations.

12.3 DEVELOPING INDUSTRIAL GRADE TOOLS

In spite of the needs for further research, the automatic learning framework is ripe for many interesting applications. Thus, in terms of software development the next stage is to design a comprehensive set of industrial grade tools for its application in real life.

Figure 12.1 depicts the envisioned overall software architecture for DSA studies by automatic learning. In this architecture the data base generation module as well as the automatic learning module will exploit trivial parallelism of their algorithms to take the best advantage of existing and future computing environments. They should be designed in a modular and "open" fashion so as to allow the easy integration of power system security analysis software and new automatic learning algorithms as soon as they become available.

With such a software platform one could apply the automatic learning framework efficiently and systematically in the planning and operating environments where security studies are presently carried out.

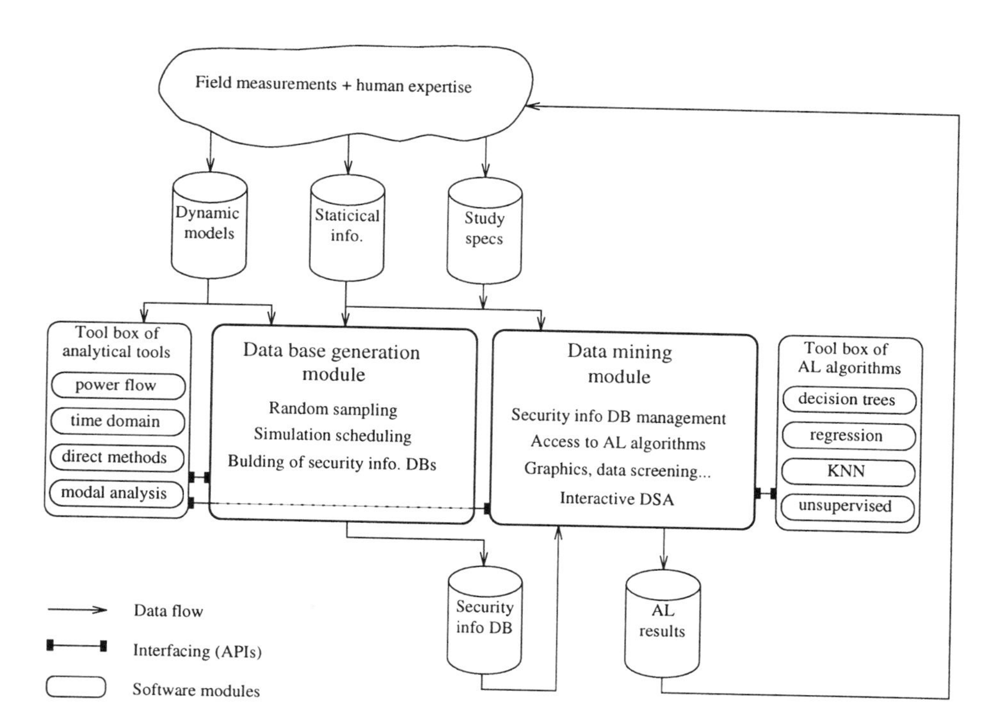

Figure 12.1. Envisioned software architecture for the application of automatic learning to DSA. Adapted from [WJ97]

III AUTOMATIC LEARNING APPLICATIONS IN POWER SYSTEMS

13 OVERVIEW OF APPLICATIONS BY TYPE

13.1 INTRODUCTION

The numerous illustrations given in Parts I and II aimed at showing many alternative ways to exploit data by automatic learning. In particular, the tool-box approach came to us as the natural way to exploit the large amounts of data within our large-scale applications to power system dynamic security assessment.

Power system security assessment has the special feature that data bases are generated by computer based simulation. This is so, because the actual experience of instabilities is - fortunately - very small for most power systems; most of them did not experience even a single blackout during the last ten years. Thus, in security assessment the engineer is responsible for exploiting numerical simulation so as to gather enough experience on plausible stability problems.

This is at the same time an advantage and a difficulty. The difficulty comes from the fact that it is necessary to be rather careful in defining the range of security simulations so as to make sure that the conclusions drawn from a study are indeed valid. The advantage is that, with enough available computing power and storage it is possible to generate as many scenarios as are deemed necessary to solve a particular task.

However, there are also many applications in power systems for which the data bases would come from the field, rather than being generated by simulation.

Many of these problems have been solved in the past by using classical techniques from approximation theory, system identification or time series modeling. Actually, these "classical" methods also belong to automatic learning and can be used in conjunction with the other methods (e.g. artificial neural networks and machine learning) presented in this book. In particular, we discuss below some of these applications where automatic learning can be used in order to take advantage from the available information.

One of the messages that we would like to convey is that engineers trying to solve such problems should look at the whole tool-box of methods, and not hesitate to combine different techniques to yield a full, practical solution. They can use clustering techniques and correlation analysis, together with decision or regression tree induction during the first steps of data mining, as well as appropriate graphical visualizations. Later on, when the problem they address has been well enough circumscribed, possibly decomposed into some simpler subproblems, they can apply either classical statistical techniques or neural network based ones, in order to exploit their data in the most efficient way.

While this is what "traditional" statisticians have often done in the past, we believe that it is now time for the engineers in charge of power systems to understand the basics of the available automatic learning methods. This would enable them to select the most appropriate way to apply these methods to their problem, on the ground of their own practical knowledge.

Indeed, our experience shows that only the conjunction of problem specific expertise and knowledge of the features of various data mining methods can lead to successful applications. Thus, probably the description of automatic learning methods given in this book will be useful to engineers which need to solve other problems. It is also hoped that the various applications described in the context of security assessment will give them some interesting ideas on how to look at their problems and data.

In the next sections we will briefly discuss four generic classes of problems encountered in power systems, and provide some indications of the possible use of automatic learning.

We do not attempt to survey publications on related works, thus we will provide only a very sparse list of bibliographical references. While our focus is on power systems, most of our comments are certainly also relevant in the context of other industrial applications of automatic learning.

At the end of this chapter we provide some indications to help the interested reader to find his way through the literature on automatic learning applications.

13.2 DESIGN

We start by discussing design problems, since they are very similar to security assessment. Design is indeed generally using numerical simulation to assess various design options.

13.2.1 Present practice

In power systems, such design problems mainly concern the appropriate *location* of some device, its *dimensioning* and the *tuning* of some of its parameters.

A typical example would be a study concerning a new static VAr compensator (SVC). This would involve identifying the best geographical location of the SVC, deciding its type and capacity, and possibly tuning some of its parameters.

Another typical application would be to tune parameters of power system stabilizers (PSSs) equipping some plants. For example, after some important changes in a system, such as the interconnection with some neighboring systems, undamped oscillations may some times appear which would imply that some PSSs must be re-tuned.

Still another design problem would be to select the number and location of new measurement devices. For example, when designing a wide area monitoring system (WAMS) using synchronized phasor measurement devices, it is necessary to decide how many such devices should be installed and in which substations they should be located. Similar problems arise when defining pilot nodes for secondary voltage control systems [LSLP89].

Present practice for solving such design problems mainly relies on human expertise. It consists in using linear system theory methods (e.g. eigenvalue computations and modal analysis) in order to first determine the system main physical characteristics for some typical operating conditions. The results of this analysis, together with the engineers' physical intuition will lead to a reduced number of design options, which are further analyzed by detailed numerical simulation in order to evaluate them more accurately, and to tune some parameters. In this approach, the quality of the resulting design would rely highly on the quality of the prior expertise available, and on the intuition of the engineer in charge of the study.

Notice that this methodology is very similar to present day practices in off-line security assessment studies. Thus, we believe that the automatic learning based framework developed for security assessment could be easily adapted to such design problems. Engineers could screen a larger number of design options, and validate them more systematically with respect to a broader set of future conditions.

13.2.2 Hypothetical application

To further illustrate how automatic learning could be used for design, let us imagine a hypothetical application in the context of the above SVC design problem.

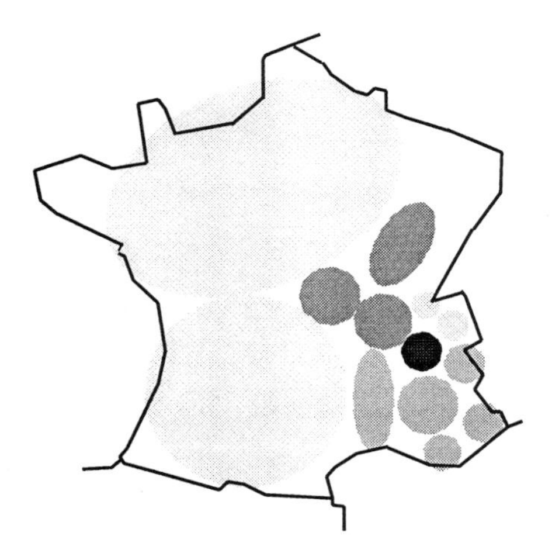

Figure 13.1. MV voltage zones

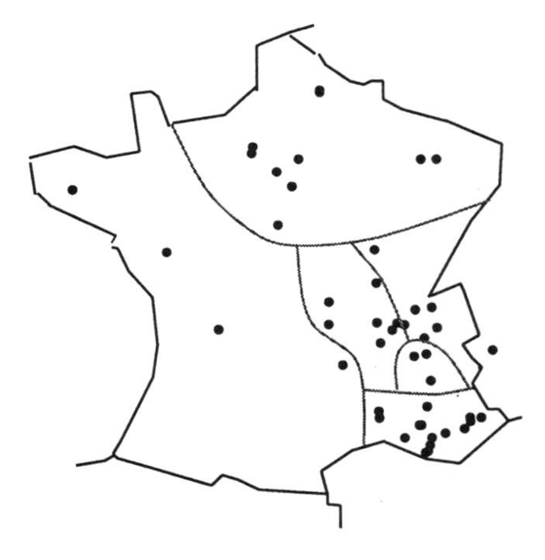

Figure 13.2. EHV voltage zones

The study could be started by the identification of the electrically "coherent" regions in the power system. A data base composed of various static and dynamic operating scenarios would be generated in order to analyze voltage coherent regions. This would lead to an application of unsupervised learning in order to analyze correlations among voltages at different nodes of the system, so as to identify substations in each region which voltage is representative of the regional behavior.

In particular, techniques such as those discussed in §§3.6.1 and 4.5.2 would be appropriate. For example, Figs. 13.1 and 13.2 illustrate such an analysis carried out in the global dynamic security study on the EDF system (see §§9.6.2 and 10.4).

Figure 13.1 shows regions of coherent MV (20kV) voltage behavior. Each ellipsoid covers a zone of load buses (notice that the zones in the South-East study region where defined with higher resolution). Ellipsoids of the same grey level are coherent and thus are classified into the same region. These regions were defined automatically by correlation analysis of the MV voltages computed on more than 1000 different simulation scenarios. The MV voltage coherent regions provide information on which kind of voltage collapse modes exist in the data base. They could for example be exploited for defining voltage collapse mitigation schemes (load shedding or on-load-tap-changer blocking).

Figure 13.2 shows regions of coherent EHV (400kV) behavior. Each bullet corresponds to an EHV bus and regions depicted on the diagram correspond to coherent variations in the data base. The interpretation is that voltage variations inside a region are highly correlated while variations of voltages in different regions are more or less independent. Such EHV regions thus indicate how the overall transmission system may react to voltage collapses.

After having identified the regions, a more refined analysis of the individual voltages in each one would allow us to identify "central" EHV buses where the overall behavior of the zone is best observed.

Having carried out this preliminary analysis, the next step in the SVC design problem could consist in generating a new data base to study the best substation to place the SVC. Random sampling would be used in order to consider a certain number of possible locations combined with a large variety of operating conditions. Each one of these scenarios could then be simulated under various assumptions and modelings (steady state, small signal, large disturbances) and the resulting information stored in a data base. Further, automatic learning could be applied in order to determine among the possible locations the most appropriate one and also to find out which parameters of the operating point would influence most strongly the behavior of the device.

Once the substation to place the SVC would be defined, a third more refined study could be carried out in order to fix various parameters. For example, several possible sizes and various tuning parameters could be screened in order to finalize the design.

It is not necessary to go further along this hypothetical example, to make clear that many of the benefits of automatic learning in security assessment would also be of value here.

In particular, a larger number of alternative design possibilities could be screened and the final choices would be more systematically validated. Furthermore, during the data base generation a large variety of possible operating conditions could be screened and various uncertain modeling parameters could be randomized, resulting in a more robust solution.

13.2.3 Design of adaptive systems

Adaptive (control or protection) systems are an alternative to robust designs in order to cope with the variety of operating conditions under which such systems must operate. The idea consists in determining which parameters influence most strongly the behavior of a control system. Then, a representative number of operating conditions can be defined and control laws designed adapted to each one. In real-time, the applicable control law is selected according to some decision rules.

Thus, automatic learning would allow one to determine the parameters to which the optimal control law of a plant is most sensitive and then learn the scheduling rules to select the appropriate control law in real-time. Let us mention in this context the possibility of using fuzzy rules, so as to interpolate smoothly among the various control laws. Notice that such hybrid systems become more and more popular, at least at the research level [HT96].

13.3 MODELING

As we have seen, most of the tasks which need to be solved in the context of power system planning and operation rely on models of the physical behavior of the system.

Some of these models may be derived with adequate accuracy from first principles together with classical system identification techniques based on appropriate measurements.

Some others are however intrinsically of approximate nature, either because of lack of information (e.g. modeling of neighboring systems) or because of intractability (e.g. modeling of load seen from the EHV system). In such circumstances, choosing an appropriate model is always difficult because of non-linearity and variability in time. However, much data can be made available by recording information from real-time measurement systems and disturbance recorders. Automatic learning will not change the nature of these problems, but it will allow to exploit these data in a systematic fashion, so as to assess the accuracy of existing models and possibly improve them.

For example, presently most Utilities use rather simple parametric models to take into account load sensitivity to voltage and frequency in simulations. Furthermore, except for some special loads (e.g. large industrial plants) the same set of parameter values is generally used for all buses and load levels. But, in the real world load behavior will vary significantly from one geographical region to another as well as a function of the load level and/or seasonal variations. The same holds true, at least to some extent, for external equivalents. Thus, we believe that making better use of the available data would indeed allow to model the load and external equivalents in a more accurate way.

In particular, clustering techniques may be useful in order to decompose a data base into homogeneous groups in terms of load behavior, so as to facilitate the identification of a dictionary of simple black-box models. The latter could be obtained using combinations of regression trees and MLPs [TK96]. Thus, using the data mining approach together with a tool-box of automatic learning methods would broaden the perspective and increase flexibility.

In addition, while extracting approximate models by automatic learning it would be possible to estimate how accurately they represent real-life, i.e. how much uncertainty remains which is not explained by the model. We could then feed back this information into the data base generation specifications for security assessment or design problems, so as to randomize the derived black-box models in a realistic way.

Another interesting modeling application consist in building non-linear state observers for local control devices. For example, reference [CYQ97] explores the possibility of using MLPs to observe remote system state variables from local measurements available in a substation where a FACTS is installed. Using this information in the FACTS control law would possibly enhance its stabilizing effect.

Notice that some of the models which need to be used in some applications are stochastic in essence. For example, in some planning studies of hydro-thermal systems, it is necessary to simulate the hydrological conditions, which are essentially of stochastic nature.

13.4 FORECASTING

Forecasting aims at predicting the future outcome of a system which is intrinsically uncertain. In power systems it is necessary to forecast the future, because decisions can not be implemented immediately and because they have to be evaluated for a certain period of "activity".

Depending on the application and on the time horizon, quantities which have to be forecasted are various : meteorological conditions, load demand, level of reservoirs, macro-economic trends, solar wind...

Global load forecasting, and specially short-term global load forecasting is one of the areas where automatic learning has already received several actual applications. In this context, various different kinds of artificial neural networks have been investigated, and the most sophisticated systems use a combination of different methods, comprising MLPs, fuzzy rules, Kohonen feature maps as well as more classical statistical models for time-series regression (see e.g. [BTKS96, BKS96]).

Another interesting area for research concerns very short term *nodal* load forecasting, to be used in enhancing real-time system monitoring and security assessment. In particular, it would be interesting to use such predicted values to enrich the information exploited by state estimators, and in particular to provide the required redundancy to detect topological errors. Further, it would be very valuable to obtain a 30-minute prediction of the individual load demands, so as to run the security assessment on the predicted state rather than on past information. A first investigation on the EHV system of Electricité de France has shown that multilayer perceptrons are promising, while clustering techniques together with regression trees may be useful for physical analysis of load behavior.

13.5 MONITORING

Monitoring is a very broad application field, ranging from the monitoring of the individual devices and plants, to sophisticated system monitoring. The purpose of monitoring is mainly to detect as early as possible when a system departs from its normal conditions. This enables one to trigger alarms in order to decide some field inspection or a detailed analysis of the potential problem. With this broad definition, monitoring includes many of the activities of a control-room, in particular, security assessment.

The main characteristic of monitoring problems is that there is generally a data base available of the normal states of the system, but no experience of the abnormal situations which one tries to detect. This is also the case of security assessment, but having argued enough that security assessment may rely on simulations of the abnormal behavior, we will restrict our focus on applications where this is not the case, i.e. applications where the sole information available is a data base of recordings of normal behavior.

Actually, the fact that only information about normal behavior is available does not preclude applying automatic learning to such data bases. To show how this is possible, let us consider an example application given in [BKS96]. It concerns the monitoring of dam deformations of hydroelectric plants.

The application was carried out in the following way. First a data base of some parameters affecting dam deformations together with the observed state of the dam was collected during a certain period, assumed to be normal. Then, using supervised learning based on multilayer perceptrons a black box model was designed relating the input parameters (time of measurement, water level, air and water temperature) to the resulting deformation. During the model validation stage, its accuracy was determined on a test set of measurements not used for learning. The resulting model is used in operation to predict dam deformations from real-time measurements. These predicted deformations are compared to actual measurements and if the difference is larger than the error bound of the model, the hypothesis that the dam is in normal operation is rejected. In short, it is concluded that something must be wrong.

Thus, the general idea of this kind of approach is to use automatic learning to identify a model of normal behavior. This model can be obtained by supervised learning in the fashion described above, by modeling correlations which hold for normal operation. It could also be derived by using unsupervised learning methods, in order to model probability distributions of states in normal operation. Then, a hypothesis testing approach is applied in order to test whether a given state can be considered to belong to the population of normal states. It consists merely of checking that the correlations determined by automatic learning are still valid.

Among the practical applications which may seem promising, let us mention power plant monitoring (burner, reactor, dam), transformer monitoring, and system oscillations monitoring. For example, using real-time measurements of high accuracy and good resolution in time (e.g. phasor measurements) one could thus monitor the power system dynamics (oscillations), and possibly detect emerging problems with some degree of anticipation.

13.6 FURTHER READING

In the preceding discussion our aim was mainly to suggest some ways to apply automatic learning to various power system problems. We did not attempt to provide a survey of the large number of publications which report on the work already done within this context. We apologize to the many researchers whose work was not mentioned.

But, in order to help the interested reader to find these references by himself, we provide below some information on various publications which would be relevant.

Some years ago CIGRE TF 39-06-06 made a survey of applications of artificial neural networks to power systems, which is summarized in [NC93]. A more recent overview of various automatic learning applications to power systems is given in the December 1996 issue of the Revue-E published by the Belgian Royal Society of Electrical Engineers [We96]. Other useful sources of information are the proceedings of various conferences, in particular ISAP (Intelligent System Applications to Power Systems), PSCC (Power System Computation Conference), CPSPP (IFAC-Cigré Symposium on Control of Power Plants and Power Systems), and journals such as the IEEE Transactions on Power Systems, EPES (Electric Power and Energy Systems) and Engineering Intelligent Systems.

Finally, we would like to draw the attention of the reader on two journals which cover automatic learning and data mining in general. The first one is "Machine Learning", which covers mainly theoretical aspects of automatic learning (in spite of its title, the journal scope comprises statistical methods and artificial neural network based ones, together with machine learning stricto sensu). The second one is "Data mining and Knowledge Discovery"; it covers methodologies and case studies related to real-world applications of automatic learning, together with other technical problems which need to be solved in the KDD process. Both journals are published by Kluwer Academic.

Other relevant journals are the IEEE Transactions on "Pattern Analysis and Machine Intelligence", "Systems, Men and Cybernetics", "Knowledge and Data Engineering", "Fuzzy Systems", and "Neural Networks".

References

The numbers in italics between parentheses appearing at the end of a bibliographical entry identify the pages where this reference is cited in the text.

[AB83] P. M. Anderson and A. Bose, *A probabilistic approach to power system stability analysis*, IEEE Trans. on Power Apparatus and Systems **102** (1983), no. 8, 2430–2439. *(Cited p. 235)*

[AKA91] D. W. Aha, D. Kibler, and M. K. Albert, *Instance-based learning algorithms*, Machine Learning **6** (1991), 37–66. *(Cited p. 100)*

[AWP+93] V.B. Akella, L. Wehenkel, M. Pavella, M. Trotignon, A. Duchamp, and B. Heilbronn, *Multicontingency decision trees for transient stability assessment*, Proc. of the 11th Power Systems Computation Conference, Aug-Sept 1993, pp. 113–119. *(Cited pp. 195, 203)*

[Bar93] A. R. Barron, *Universal approximation bounds for superpositions of a sigmoidal function*, IEEE Trans. on Info. Theory **39** (1993), no. 3, 930–945. *(Cited p. 87)*

[BFOS84] L. Breiman, J. H. Friedman, R. A. Olshen, and C. J. Stone, *Classification and regression trees*, Wadsworth International (California), 1984. *(Cited pp. 6, 39, 43, 100, 105, 110, 112, 138, 143)*

[BK80] R. Billinton and P. R. S. Kuruganty, *A probabilistic index for transient stability*, IEEE Trans. on Power Apparatus and Systems **99** (1980), no. 1, 195–206. *(Cited p. 235)*

[BK88] I. Bratko and I. Kononenko, *Learning diagnostic rules from incomplete and noisy data*, AI and Statistics (B. Phelphs, ed.), Technical Press, 1988. *(Cited p. 100)*

[BKS96] T. Bauman, B. Knapp, and G. Schellstede, *Automatic learning appliactions to power system operation and control*, Revue E - SRBE - Special Issue on Automatic learning applied to power systems (1996). *(Cited pp. 253, 254)*

[BMVY93] B. Bouchon-Meunier, L. Valverde, and R. Yager (eds.), *Uncertainty in intelligent systems*, North-Holland, 1993. *(Cited p. 100)*

[BTKS96] A. G. Bakirtzis, J. B. Theocharis, S. J. Kiartzis, and K. J. Satsios, *Short term load forecasting using fuzzy neural networks*, IEEE Transactions on Power Systems (1996). *(Cited p. 253)*

[Bun92] W. L. Buntine, *Learning classification trees*, Statistics and Computing **2** (1992), 63–73. *(Cited pp. 43, 44, 100, 115)*

[BW91] W. L. Buntine and A. S. Weigend, *Bayesian back-propagation*, Complex Systems **5** (1991), 603–643. *(Cited pp. 43, 44)*

[BW95] X. Boyen and L. Wehenkel, *Fuzzy decision tree induction for power system security assessment*, Proc. of SIPOWER'95, 2nd IFAC Symp. on Control of Power Plants and Power Systems (Mexico), December 1995, pp. 151–156. *(Cited pp. 140, 219)*

[BW96] X. Boyen and L. Wehenkel, *Automatic induction of continuous decision trees*, Proc. of IPMU'96, Information Processing and Management of Uncertainty in Knowledge-Based Systems (Granada), July 1996, pp. 419–424. *(Cited p. 140)*

[CC87] C. Carter and J. Catlett, *Assessing credit card applications using machine learning*, IEEE Expert **Fall** (1987), 71–79. *(Cited p. 114)*

[Che85] P. Cheeseman, *In defense of probability*, Proc. of the IJCAI-85, 1985, pp. 1002–1009. *(Cited p. 100)*

[Cho88] P. A. Chou, *Application of information theory to pattern recognition and the design of decision trees and trellises*, Ph.D. thesis, Stanford University, June 1988. *(Cited p. 115)*

[Cho91] P. A. Chou, *Optimal partitioning for classification and regression trees*, IEEE Trans. on Pattern Analysis and Machine Intelligence **PAMI-13** (1991), no. 14, 340–354. *(Cited pp. 37, 105)*

[CM97] TF 38.02.13 CIGRE and B. Meyer, *New trends and requirements for dynamic security assessment*, To be submitted to ELECTRA (1997). *(Cited p. 172)*

[CN89] P. Clark and T. Niblett, *The CN2 induction algorithm*, Machine Learning **3** (1989), 261–283. *(Cited p. 100)*

[Coh95] Paul R. Cohen, *Empirical methods in artificial intelligence*, MIT Press, 1995. *(Cited p. 45)*

[CS91] S. Cost and S. Salzberg, *A weighted nearest neighbor algorithm for learning with symbolic features*, Tech. report, Dept. of Computer Science, John Hopkins University, 1991. *(Cited p. 100)*

[CSKS88] P. Cheeseman, M. Self, J. Kelly, and J. Stutz, *Bayesian classification*, Proc. of the 7th AAAI Conf., 1988, pp. 607–611. *(Cited p. 69)*

[Cyb89] G. Cybenko, *Approximations by superpositions of a sigmoidal function*, Matt. Contro. Signals, Syst. **2** (1989), 303–314. *(Cited p. 87)*

[CYQ97] Shen Chen, Sun Yuanzhang, and Lu Qiang, *Studies on the possibility of applying artifical neural networks to estimate real time operation parameters of power systems*, Proc. of CPSPP'97, Ifac-Cigré Symposium on Control of Power Systems and Power Plants, 1997, pp. 334–338. *(Cited p. 252)*

[Dar70] Z. Daróczy, *Generalized informationfunctions*, Information and Control **16** (1970), 36–51. *(Cited p. 123)*

[De96] T. Dillon and D. Niebur (eds), *Neural net applications in power systems*, CRL Publishing Ltd., Leics, UK, 1996. *(Cited p. 3)*

[DH73] R. O. Duda and P. E. Hart, *Pattern classification and scene analysis*, John Wiley and Sons, 1973. *(Cited pp. 48, 51, 62, 68, 92, 112)*

[Die96] T. G. Dietterich, *Statistical tests for comparing supervised classification learning algorithms*, Tech. report, Oregon State University, ftp://ftp.cs.orst.edu/pub/tgd/papers/stats.ps.gz, 1996. *(Cited p. 40)*

[Dil91] T.S Dillon, *Artificial neural network applications to power systems and their relationship to symbolic methods*, Int. J. of Elec. Power and Energy Syst. **13** (1991), no. 2, 66–72. *(Cited p. 3)*

[DK82] P. A. Devijver and J. Kittler, *Pattern recognition : A statistical approach*, Prentice-Hall International, 1982. *(Cited pp. 38, 39, 57, 58, 59, 62, 67, 68, 146)*

[DL97] T. E. Dy-Liacco, *Enhancing power system security control*, IEEE Computer Applications in Power **10** (1997), no. 3. *(Cited p. 172)*

[DM86] J. C. Dodu and A. Merlin, *New probabilistic approach taking into account reliability and operation security in EHV power system planning at EDF*, IEEE Trans. on Power Syst. **PWRS-1** (1986), 175–181. *(Cited p. 241)*

[DP78] D. Dubois and H. Prade, *Fuzzy sets and systems. Theory and applications*, Academic Press, 1978. *(Cited p. 140)*

[Dy 68] T. E. Dy Liacco, *Control of power systems via the multi-level concept*, Ph.D. thesis, Sys. Res. Center, Case Western Reserve Univ., 1968, Rep. SRC-68-19. *(Cited pp. 2, 163)*

[EGH+92] E. Euxibie, M. Goubin, B. Heilbronn, L. Wehenkel, Y. Xue, T. Van Cutsem, and M. Pavella, *Prospects of application to the french system of fast methods for transient stability and voltage security assessment*, CIGRE Report 38-208, Paris, Aug.-Sept. 1992. *(Cited p. 200)*

[ESMA+89] M. A. El-Sharkawi, R. J. Marks II, M. E. Aggoune, D. C. Park, M. J. Damborg, and L. E. Atlas, *Dynamic security assessment of power systems using back error propagation artificial neural networks*, Proc. of the 2nd Symposium on Expert Systems Application to power systems, 1989, pp. 366–370. *(Cited p. 3)*

[ESN96] M. El-Sharkawi and D. Niebur, *Tutorial course on applications of artificial neural networks to power systems*, vol. IEEE Catalog Number 96 TP 112-0, IEEE- Power Engineering Society, 1996. *(Cited p. 3)*

[FC78] L. H. Fink and K. Carlsen, *Operating under stress and strain*, IEEE Spectrum **15** (1978), 48–53. *(Cited p. 166)*

[FD92] M. Fombellida and J. Destiné, *Méthodes heuristiques et méthodes d'optimisation non contraintes pour l'apprentissage des perceptrons multicouches*, Proc. of NEURO-NIMES 92, Fifth International Conference on Neural Networks and their Applications, 1992. *(Cited pp. 85, 86)*

[Fis36] R. A. Fisher, *The use of multiple measurements in taxonomic problems*, Ann. Eugenics **7** (1936), no. Part II, 179–188. *(Cited p. 50)*

[FKCR89] R. Fischl, M. Kam, J.-C. Chow, and S. Ricciardi, *Screening power system contingencies using back propagation trained multi-perceptrons*, Proc. of the IEEE Int. Symposium on Circuits and Systems, 1989, pp. 486–494. *(Cited p. 3)*

[FL90] F. C. Fahlman and C. Lebière, *The cascaded-correlation learning architecture*, Advances in Neural Information Processing Systems II (D. S. Touretzky, ed.), Morgan Kaufmann, 1990, pp. 524–532. *(Cited p. 56)*

[FPSSU96] U. M. Fayyad, G. Piatetsky-Shapiro, P. Smyth, and R. Uthurusamy, *Advances in knowledge discovery and data mining*, AAAI Press/MIT Press, 1996. *(Cited pp. 2, 24)*

[Fri77] J. H. Friedman, *A recursive partitioning decision rule for nonparametric classification*, IEEE Trans. on Computers **C-26** (1977), 404–408. *(Cited pp. 109, 112)*

[Fri91] J. H. Friedman, *Multivariate adaptive regression splines*, The Annals of Statistics **19** (1991), no. 1, 1–141. *(Cited p. 11)*

[Fri97] J. H. Friedman, *On bias, variance, 0/1-loss, and the curse-of-dimensionality*, Data Mining and Knowledge Discovery **1** (1997), no. 1, 55–77. *(Cited p. 63)*

[FS81] J. H. Friedman and W. Stuetzle, *Projection pursuit regression*, Jour. of the Am. Stat. Ass. **76** (1981), no. 376, 817–823. *(Cited pp. 11, 56)*

[Gau26] K. F. Gauss, *Theoria combinationis observatorionum errroribus minimis obnoxiae*, Dietrich, Göttingen, 1826. *(Cited p. 1)*

[GEA77] C. L. Gupta and A. H. El-Abiad, *Transient security assessment of power systems by pattern recognition - a pragmatic approach*, Proc. IFAC Symp. on Automatic Control and Protection of power systems, 1977. *(Cited p. 2)*

[GJP95] F. Girosi, M. Jones, and T. Poggio, *Regularization theory and neural networks architectures*, Neural Computation **7** (1995), 219–269. *(Cited p. 43)*

[Gol89] D. E. Goldberg, *Genetic algorithms in search, optimization, and machine learning*, Adisson Wesley, 1989. *(Cited p. 155)*

[Han81] D. J. Hand, *Discrimination and classification*, John Wiley and Sons, 1981. *(Cited pp. 62, 68)*

[Hay94] S. Haykin, *Neural networks. A comprehensive foundation*, IEEE Press, 1994. *(Cited pp. 43, 78, 89)*

[HCS94] N. D. Hatziargyriou, G. C. Contaxis, and N. C. Sideris, *A decision tree method applied to on-line steady-state security assessment*, IEEE Trans. on Power Syst. **9** (1994), no. 2, 1052–1061. *(Cited p. 3)*

[HF69] E. G. Henrichon and K. S. Fu, *A non-parametric partitioning procedure for pattern classification*, IEEE Trans. on Computers (1969), no. 7, 614–624. *(Cited pp. 100, 109)*

[HKP91] J. Hertz, A. Krogh, and R. G. Palmer, *Introduction to the theory of neural computation*, Addison Wesley, 1991. *(Cited pp. 85, 89, 90, 93)*

[HMS66] E. B. Hunt, J. Marin, and P. J. Stone, *Experiments in induction*, Wiley, 1966. *(Cited pp. 1, 100)*

[HP91] E. Handschin and A. Petroianu, *Energy management systems*, Springer-Verlag, 1991. *(Cited p. 169)*

[HSW89] K. Hornik, M. Stinchcombe, and H. White, *Multilayer feedforward networks are universal approximators*, Neural Networks **2** (1989), no. 5, 359–366. *(Cited p. 87)*

[HT96] P. Hoang and K. Tomsovic, *Design and analysis of an adaptive fuzzy power system stabilizer*, IEEE Trans. on Power Syst. (1996). *(Cited p. 251)*

[HTL+90] Y. Harmand, M. Trotignon, J. F. Lesigne, J. M. Tesseron, C. Lemaître, and F. Bourgin, *Analyse d'un cas d'écroulement en tension et proposition d'une philosophie de parades fondées sur des horizons temporels différents*, CIGRE Report 38/39-02, Paris, August 1990. *(Cited p. 208)*

[HWP95] I. Houben, L. Wehenkel, and M. Pavella, *Coupling of K-NN with decision trees for power system transient stability assessment*, IEEE Conference on Control Applications (Albany (NJ)), 1995, pp. 825–832. *(Cited p. 155)*

[HWP97a] I. Houben, L. Wehenkel, and M. Pavella, *Genetic algorithm based k nearest neighbors*, Proc. CIS-97, IFAC Conf. on Contr. of Indust. Syst. (Belfort, Fr), 1997. *(Cited p. 155)*

[HWP97b] Isabelle Houben, Louis Wehenkel, and Mania Pavella, *Hybrid adaptive nearest neighbor approaches to dynamic security assessment*, Proc. of CPSPP'97 (Beijing), International Academic Publishers, 1997, pp. 685–690. *(Cited pp. 155, 187)*

[HYLJ93] J. N. Hwang, S. S. You, S. R. Lay, and I. C. Jou, *What's wrong with a cascaded correlation learnign network : a projection pursuit learning perspective*, Tech. report, Info. Proc. Lab., Dep.t of Elec. Eng., University of Washington, September 1993. *(Cited p. 56)*

[Jay95] E. T. Jaynes, *Probability theory: the logic of science*, fragmentary edition of july 1995 ed., edited by E. T. Jaynes, available from http://omega.albany.edu:8008/JaynesBook.html, 1995. *(Cited p. 33)*

[Jen96] F. V. Jensen, *An introduction to Bayesian networks*, UCL Press, 1996. *(Cited p. 100)*

[JWVP95] Y. Jacquemart, L. Wehenkel, T. Van Cutsem, and P. Pruvot, *Statistical approaches to dynamic security assessment: The data base generation problem*, Proc. of SIPOWER'95, 2nd IFAC Symp. on Control of Power Plants and Power Systems, December 1995, pp. 243–246. *(Cited pp. 178, 211, 213, 214, 216)*

[KAG96] T. Kostic, J. J. Alba, and A. J. Germond, *Optimization and learning of load restoration strategies*, Proc. of PSCC'96 (Dresden), 1996. *(Cited p. 167)*

[KBR84] I. Kononenko, I. Bratko, and E. Roskar, *Experiments in automatic learning of medical diagnosis rules*, Tech. report, Jozef Stefan Institute, 1984. *(Cited pp. 100, 110)*

[KD91] J. D. Kelly and L. Davis, *Hybridizing the genetic algorithm and the k nearest neighbors classification algorithm*, Proceedings of the fourth international conference on Genetic Algorithms-Morgan Kaufmann Publishers, 1991, pp. 377–383. *(Cited p. 155)*

[KL95] R. Kohavi and Chia-Hsin Li, *Oblivious decision trees, graphs and top-down pruning*, Proc. of IJCAI'95, Morgan-Kaufmann, 1995, pp. 1071–1077. *(Cited p. 115)*

[KM97] P. Kundur and G. K. Morisson, *A review of definitions and classification of stability problems in today's power systems*, Panel session on Stability Terms and Definitions, IEEE PES Winter Meeting (1997). *(Cited p. 167)*

[KO95] J. R. Koehler and A. B. Owen, *Computer experiments*, Tech. report, Standford University - Department of Statistics, 1995, http://playfair.Stanford.EDU/reports/owen/. *(Cited p. 179)*

[Koh90] T. Kohonen, *The self-organizing map*, Proceedings of the IEEE **78** (1990), no. 9, 1464–1480. *(Cited pp. 18, 90, 91, 97)*

[Koh95] T. Kohonen, *Self-organizing maps*, Springer Verlag, 1995. *(Cited p. 91)*

[Kun94] Prabha Kundur, *Power system stability and control*, McGraw-Hill, 1994. *(Cited p. 167)*

[Kvå87] T. O. Kvålseth, *Entropy and correlation: some comments*, IEEE Trans. on Systems, Man and Cybernetics **SMC-17** (1987), no. 3, 517–519. *(Cited pp. 124, 126, 130)*

[Lap10] P. S. Laplace, *Mémoire sur les approximations des formules qui sont des fonctions de très grands nombres et sur leur application aux probablités*, Mémoires de l'Académie des Sciences de Paris, 1810. *(Cited p. 1)*

[LSLP89] P. Lagonotte, J. C. Sabonnadière, J. Y. Léost, and J. P. Paul, *Structural analysis of the electrical system : application to the secondary voltage control in France*, IEEE Trans. on Power Syst. **PWRS-4** (1989), no. 4, 479–486. *(Cited p. 249)*

[LWCC97] C. Lebrevelec, L. Wehenkel, P. Cholley, and J. P. Clerfeuille, *A probabilistic approach to improve security assessment*, Proc. IERE Workshop on Future Directions in Power System Reliability (Palo-Alto), 1997, pp. 243–252. *(Cited p. 235)*

[LWW+89] C.-C. Liu, S.-M. Wang, L. Wong, H. Y. Marathe, and M. G. Lauby, *A self learning expert system for voltage control of power systems*, Proc. 2nd Symp. on Expert Systems Application to Power Systems, 1989, pp. 462–468. *(Cited p. 106)*

[MD94] S. Muggleton and L. De Raedt, *Inductive logic programming : theory and methods*, Journal of Logic Programming **19,20** (1994), 629–679. *(Cited p. 46)*

[MFV+97] J. D. McCalley, A. A. Fouad, V. Vittal, A. A. Irizarry-Rivera, B. L. Agrawal, and R. G. Farmer, *A risk-based security index for determining operating limits in stability limited electric power systems*, 1997, IEEE PES 96 SM 505-6. *(Cited p. 235)*

[Mic83] R. S. Michalski, *A theory and methodology of inductive learning*, Artificial Intelligence **20** (1983), 111–161. *(Cited p. 100)*

[Min89] J. Mingers, *An empirical comparison of pruning methods for decision tree induction*, Machine Learning **4** (1989), 227–243. *(Cited p. 110)*

[MK95] J. D. McCalley and B. A. Krause, *Rapid transmission capacity margin determination for dynamic security assessment using artificial neural networks*, Electric Power Systems Research (1995). *(Cited p. 3)*

[MKSB93] S. Murthy, S. Kasif, S. Salzberg, and R. Beigel, *OC1 : randomized induction of oblique trees*, Proc. of the AAAI-93, 1993. *(Cited p. 112)*

[MP43] W. S. McCulloch and W. Pitts, *A logical calculus of ideas immanent in nervous activity*, Bulletin of Mathematical Biophysics **5** (1943), 115–133. *(Cited pp. 1, 71)*

[MP69] M. L. Minsky and S. A. Papert, *Perceptrons*, MIT Press, 1969. *(Cited p. 71)*

[MS63] J. N. Morgan and J. A. Sonquist, *Problems in the analysis of survey data, and a proposal*, J. of the Amer. Stat. Ass. **58** (1963), 415–434. *(Cited p. 100)*

[MS92] B. Meyer and M. Stubbe, *EUROSTAG, a single tool for power system simulation*, Transmission and Distribution International **3** (1992), no. 1, 47–52. *(Cited p. 191)*

[MST94] D. Michie, D. J. Spiegelhalter, and C. C. Taylor (eds.), *Machine learning, neural and statistical classification*, Ellis Horwood, 1994, Final rep. of ESPRIT project 5170 - StatLog. *(Cited pp. 24, 57, 93, 94, 97, 100, 201, 202, 228)*

[MT91] H. Mori and Y. Tamura, *An artificial neural-net based approach to power system voltage stability*, Proc. of the 2nd Int. Workshop on Bulk Power System Voltage Phenomena - Voltage Stability and Security, August 1991, pp. 347–358. *(Cited pp. 3, 91, 93)*

[MW97] R. S. Michalski and J. Wnek (eds.), *Machine learning - special issue on multistrategy learning*, vol. 27, Kluwer Academic, 1997. *(Cited p. 149)*

[NC93] D. Niebur and CIGRE TF 38-06-06, *Artificial neural networks for power systems: a literature survey*, Engineering Intelligent Systems for Electrical Engineering and Communications **1** (1993), no. 3, 133–158. *(Cited p. 255)*

[NG91] D. Niebur and A. Germond, *Power system static security assessment using the Kohonen neural network classifier*, Proc. of the IEEE Power Industry Computer Application Conference, May 1991, pp. 270–277. *(Cited pp. 3, 91, 93)*

[OH91] D. R. Ostojic and G. T. Heydt, *Transient stability assessment by pattern recognition in the frequency domain*, IEEE Trans. on Power Syst. **PWRS-6** (1991), no. 1, 231–237. *(Cited p. 3)*

[Pao89] Y. H. Pao, *Adaptive pattern recognition and neural networks*, Addison-Wesley, 1989. *(Cited p. 90)*

[Par87] D. B. Parker, *Optimal algorithms for adaptive networks : second order back propagation, second order direct propagation, second order Hebbian learning*, Proc. of IEEE First Int. Conf. on Neural Networks, 1987, pp. 593–600. *(Cited p. 85)*

[PDB85] Y. H. Pao, T. E. Dy Liacco, and I. Bozma, *Acquiring a qualitative understanding of system behavior through AI inductive inference*, Proc.

of the IFAC Symp. on Electric Energy Systems, 1985, pp. 35–41. *(Cited pp. 3, 164)*

[Pea88] J. Pearl, *Probabilistic reasoning in intelligent systems - Networks of plausible inference*, Morgan-Kaufman, 1988. *(Cited p. 100)*

[PM93] M. Pavella and P.G. Murthy, *Transient stability of power systems : theory and practice*, J. Wiley, 1993. *(Cited p. 168)*

[PMOP92] M. V. F. Pereira, M. E. P. Maceira, G. C. Oliveira, and L. M. V. G. Pinto, *Combining analytical models and monte-carlo techniques in probabilistic power system analysis*, IEEE Trans. on Power Syst. **PWRS-7** (1992), 265–272. *(Cited p. 240)*

[PPEAK74] C. K. Pang, F. S. Prabhakara, A. H. El-Abiad, and A. J. Koivo, *Security evaluation in power systems using pattern recognition*, IEEE Trans. on Power Apparatus and Systems **93** (1974), no. 3. *(Cited p. 2)*

[Qui83] J. R. Quinlan, *Learning efficient classification procedures and their application to chess endgames*, Machine Learning : An artificial intelligence approach. (R. S. Michalski, J. Carbonell, and T. Mitchell, eds.), Morgan Kaufman, 1983, pp. 463–482. *(Cited pp. 100, 106, 194)*

[Qui86a] J. R. Quinlan, *The effect of noise on concept learning*, Machine Learning II. (R. S. Michalski, J. Carbonell, and T. Mitchell, eds.), Morgan Kaufman, 1986, pp. 149–166. *(Cited pp. 110, 130)*

[Qui86b] J. R. Quinlan, *Induction of decision trees*, Machine Learning **1** (1986), 81–106. *(Cited p. 114)*

[Qui87] J. R. Quinlan, *Simplifying decision trees*, Int. J. of Man-Mach. Studies **27** (1987), 221–234. *(Cited p. 110)*

[Qui93] J. R. Quinlan, *C4.5. programs for machine learning*, Morgan Kaufman, 1993. *(Cited p. 6)*

[RHW86] D. E. Rumelhart, G. E. Hinton, and R. J. Williams, *Learning representations by back-propagating errors*, Nature **323** (1986), 533–536. *(Cited p. 72)*

[Ris78] J. Rissanen, *Modeling by shortest data description*, Automatica **14** (1978), 465–471. *(Cited p. 45)*

[RKTB94] S. Rovnyak, S. Kretsinger, J. Thorp, and D. Brown, *Decision trees for real-time transient stability prediction*, IEEE Trans. on Power Syst. **9** (1994), no. 3, 1417–1426. *(Cited pp. 3, 172)*

[RL91] M. D. Richard and R. P. Lippmann, *Neural network classifiers estimate Bayesian a posteriori probabilities*, Neural Computation **3** (1991), 461–483. *(Cited p. 84)*

[RN95] S. J. Russel and P. Norvig, *Artificial intelligence - A modern approach*, Prentice-Hall, 1995. *(Cited p. 143)*

[Ros63] F. Rosenblatt, *Principles of neurodynamics*, Spartan, 1963. *(Cited p. 71)*

[Rou80] E. M. Rounds, *A combined nonparametric approach to feature selection and binary decision tree design*, Pattern Recognition **12** (1980), 313–317. *(Cited p. 110)*

[Rub81] R. Y. Rubinstein, *Simulation and the Monte Carlo method*, J. Wiley, 1981. *(Cited p. 239)*

[Sal91] S. Salzberg, *A nearest hyperrectangle learning method*, Machine Learning **6** (1991), 251–276. *(Cited p. 100)*

[Set90] I. K. Sethi, *Entropy nets : from decision trees to neural networks*, Proceedings of the IEEE **78** (1990), no. 10, 1605–1613. *(Cited p. 151)*

[Sha90] S. C. Shapiro, *Encyclopedia of artificial intelligence*, Wiley-Interscience, 1990. *(Cited p. 99)*

[Shr66] Yu. A. Shrieder, *The Monte Carlo method. The method of statistical trials*, Ed.) (Pergamon, 1966. *(Cited p. 240)*

[SP89] D. J. Sobajic and Y.H. Pao, *Artificial neural-net based dynamic security assessment for electric power systems*, IEEE Trans. on Power Syst. **PWRS-4** (1989), no. 4, 220–228. *(Cited p. 3)*

[SP97] P. W. Sauer and M. A. Pai, *Power system dynamics and stability*, Prentice Hall Engineering, 1997. *(Cited p. 167)*

[SW86] C. Stanfill and D. Waltz, *Toward memory-based reasoning*, Communications of the ACM **29** (1986), no. 12, 1213–1228. *(Cited p. 100)*

[SW96] E. F. Sánchez-Úbeda and L. Wehenkel, *The hinges model : a one-dimensional piecewise linear model*, Tech. report, University of Liège, September 1996, 24 pages. *(Cited p. 56)*

[Tay93] C. W. Taylor, *Power system voltage stability*, McGraw-Hill, 1993. *(Cited p. 168)*

[TK96] R. J. Thomas and Bih-Yuan Ku, *A neural network representation of dynamic loads for calculating security limits in electrical power systems*, Revue E - SRBE (1996). *(Cited p. 252)*

[Tou74] G. T. Toussaint, *Bibliography on estimation of misclassification*, IEEE Trans. on Information Theory **IT-20** (1974), no. 4, 472–479. *(Cited p. 38)*

[UPK+97] K. Uhlen, A. Pettersteig, G. H. Kjolle, G. G. Lovas, and M. Meisingset, *On-line security assessment and control - probabilistic versus deterministic operational criteria*, Proc. IERE Workshop on Future Directions in Power System Reliability (Palo-Alto), 1997, pp. 169–178. *(Cited p. 235)*

[Utg88] P. E. Utgoff, *Perceptron trees : a case study in hybrid concept representation*, AAAI-88. Proc. of the 7th Nat. Conf. on Artificial Intelligence, Morgan Kaufman, 1988, pp. 601–606. *(Cited p. 112)*

[Val84] L. G. Valiant, *A theory of the learnable*, Communications of the ACM **27** (1984), no. 11, 1134–1142. *(Cited p. 44)*

[Vap95] V. N. Vapnik, *The nature of statistical learning theory*, Springer Verlag, 1995. *(Cited pp. 26, 42, 44)*

[VJMP96] T. Van Cutsem, Y. Jacquemart, J.N. Marquet, and P. Pruvot, *A comprehensive analysis of mid-term voltage stability*, IEEE Trans. on Power Syst. (1996). *(Cited p. 212)*

[VL86] M. Vincelette and D. Landry, *Stability limit selection of the Hydro-Québec power system : a new software philosophy*, Proc. of the 2nd Int. IEE Conf. on Power Syst. Monitoring and Control, 1986, pp. 367–371. *(Cited p. 117)*

[VPGN94] B. X. Vieira Filho, M. V. F. Pereira, P. Gomes, and E. Nery, *A probabilistic approach to determine the proximity of the voltage collapse region*, CIGRE, paper 38-201 (1994). *(Cited p. 235)*

[VPLH96] H. Vu, P. Pruvot, C. Launay, and Y. Harmand, *An improved voltage control on large scale power systems*, IEEE Trans. on Power Syst. **11** (1996), no. 3, 1295–1303. *(Cited p. 208)*

[VWM+97] V. Van Acker, S. Wang, J. McCalley, G. Zhou, and M. Mitchell, *Data generation using automated security assessment for neural network training*, Proc. of North American power Conference, 1997. *(Cited p. 179)*

[WA93] L. Wehenkel and V.B. Akella, *A hybrid decision tree - neural network approach for power system dynamic security assessment*, Proc. of the

4th Int. Symp. on Expert Systems Application to Power Systems (Melbourne, Australia), January 1993, pp. 285–291. *(Cited pp. 151, 152, 195)*

[Wat87] R. L. Watrous, *Learning algorithms for connectionist networks : applied gradient methods of nonlinear optimization*, Proc. of IEEE First Int. Conf. on Neural Networks, 1987, pp. 619–627. *(Cited p. 85)*

[We96] L. Wehenkel and M. Pavella (eds), *Special issue on automatic learning applications to power systems of the Revue-E*, SRBE - Belgium, December 1996. *(Cited p. 255)*

[Weh90] L. Wehenkel, *Une approche de l'intelligence artificielle appliquée à l'évaluation de la stabilité transitoire des réseaux électriques*, Ph.D. thesis, University of Liège, May 1990, In French. *(Cited pp. 119, 123, 131)*

[Weh92a] L. Wehenkel, *An information quality based decision tree pruning method*, Proc. of the 4th Int. Congr. on Information Processing and Management of Uncertainty in Knowledge based Systems - IPMU'92, July 1992, pp. 581–584. *(Cited pp. 119, 123)*

[Weh92b] L. Wehenkel, *A probabilistic framework for the induction of decision trees*, Tech. report, University of Liège, 1992. *(Cited p. 119)*

[Weh93a] L. Wehenkel, *Decision tree pruning using an additive information quality measure*, Uncertainty in Intelligent Systems (B. Bouchon-Meunier, L. Valverde, and R.R. Yager, eds.), Elsevier - North Holland, 1993, pp. 397–411. *(Cited pp. 43, 110, 115, 119, 123, 132)*

[Weh93b] L. Wehenkel, *Evaluation de la sécurité en temps réel: approche par arbres de décision*, Actes de la journée d'études SEE, *Intégration des techniques de l'intelligence artificielle dans la conduite et la gestion des réseaux électriques*, March 1993, pp. 11–20. *(Cited p. 205)*

[Weh94] L. Wehenkel, *A quality measure of decision trees. interpretations, justifications, extensions*, Tech. report, University of Liège - Belgium, 1994. *(Cited pp. 70, 119)*

[Weh95a] L. Wehenkel, *Machine learning approaches to power system security assessment*, Faculty of Applied Sciences - University of Liège, No 142, 400 pages, 1995. *(Cited pp. 60, 117, 169, 195)*

[Weh95b] L. Wehenkel, *A statistical approach to the identification of electrical regions in power systems*, Proc. of Stockholm Power Tech (IEEE paper # SPT PS 17-03-0237), June 1995, pp. 530–535. *(Cited p. 96)*

[Weh96a] L. Wehenkel, *Contingency severity assessment for voltage security using non-parametric regression techniques*, IEEE Trans. on Power Syst. **PWRS-11** (1996), no. 1, 101–111. *(Cited pp. 138, 221, 228)*

[Weh96b] L. Wehenkel, *On uncertainty measures used for decision tree induction*, Proc. of IPMU'96, Information Processing and Management of Uncertainty in Knowledge-Based Systems (Granada), July 1996, pp. 413–418. *(Cited pp. 123, 124)*

[Weh97a] L. Wehenkel, *Discretization of continuous attributes for supervised learning. Variance evaluation and variance reduction*, Proc. of IFSA'97, Int. Fuzzy Systems Assoc. World Congress. Special session on Learning in a fuzzy framework, 1997. *(Cited pp. 43, 115, 127, 140)*

[Weh97b] L. Wehenkel, *Machine learning approaches to power-system security assessment*, IEEE Expert, Intelligent Systems & their Applications **12** (1997), no. 3. *(Cited p. 240)*

[WHP95a] L. Wehenkel, I. Houben, and M. Pavella, *Automatic learning approaches for on-line transient stability preventive control of the Hydro-Québec system - Part II. A tool box combining decision trees with neural nets and nearest neighbor classifiers optimized by genetic algorithms*, Proc. of SIPOWER'95, 2nd IFAC Symp. on Control of Power Plants and Power Systems, December 1995, pp. 237–242. *(Cited pp. 186, 187)*

[WHP+95b] L. Wehenkel, I. Houben, M. Pavella, L. Riverin, and G. Versailles, *Automatic learning approaches for on-line transient stability preventive control of the Hydro-Québec system - Part I. Decision tree approaches*, Proc. of SIPOWER'95, 2nd IFAC Symp. on Control of Power Plants and Power Systems, December 1995, pp. 231–236. *(Cited pp. 186, 187)*

[WJ95] L. Wehenkel and Y. Jacquemart, *Use of Kohonen feature maps for the analysis of voltage security related electrical distances*, Proc. of ICANN'95, Int. Conf. on Artificial Neural Networks, NEURONîMES'95 (Industrial conference), October 1995, pp. 8.3.1–8.3.7. *(Cited pp. 91, 223)*

[WJ97] L. Wehenkel and Y. Jacquemart, *Automatic learning methods - Application to dynamic security assessment*, Tutorial course given at IEEE PICA'97, 1997. *(Cited p. 243)*

[WK91] S. M. Weiss and C. A. Kulikowski, *Computer systems that learn*, Morgan Kaufmann, USA, 1991. *(Cited pp. 38, 40, 100)*

[WLTB97a] L. Wehenkel, C. Lebrevelec, M. Trotignon, and J. Batut, *A probabilistic approach to the design of power systems protection schemes against blackouts*, Proc. IFAC-CIGRE Symp. on Control of Power Plants and Power Systems (Beijing), 1997, pp. 506–511. *(Cited pp. 171, 181, 188, 189, 191)*

[WLTB97b] L. Wehenkel, C. Lebrevelec, M. Trotignon, and J. Batut, *A step towards probabilistic global dynamic security assessment*, Submitted (1997). *(Cited p. 192)*

[Wol94] D. H. Wolpert (ed.), *The mathematics of generalization*, Addison Wesley, 1994, Proc. of the SFI/CNLS Workshop on Formal Approaches to Supervised Learning. *(Cited pp. 1, 42, 45)*

[WP91] L. Wehenkel and M. Pavella, *Decision trees and transient stability of electric power systems*, Automatica **27** (1991), no. 1, 115–134. *(Cited pp. 115, 194)*

[WP93a] L. Wehenkel and M. Pavella, *Advances in decision trees applied to power system security assessment*, Proc. of APSCOM-93, IEE Int. conf. on advances in power system Control, Operation and Management (Invited), December 1993, pp. 47–53. *(Cited p. 170)*

[WP93b] L. Wehenkel and M. Pavella, *Decision tree approach to power system security assessment*, Int. J. of Elec. Power and Energy Syst. **15** (1993), no. 1, 13–36. *(Cited p. 164)*

[WP96] L. Wehenkel and M. Pavella, *Why and which automatic learning approaches to power systems security assessment*, Proc. of CESA'96, IMACS/IEEE SMC Multiconference on Computational Engineering in Systems Applications (Lille, Fr), July 1996, pp. 1072–1077. *(Cited p. 156)*

[WPEH94] L. Wehenkel, M. Pavella, E. Euxibie, and B. Heilbronn, *Decision tree based transient stability method - a case study*, IEEE Trans. on Power Syst. **PWRS-9** (1994), no. 1, 459–469. *(Cited pp. 141, 195, 197, 206)*

[WTY88] F. F. Wu, Y. K. Tsai, and Y. X. Yu, *Probabilistic steady state and dynamic security assessment*, IEEE Trans. on Power Systems **3** (1988), no. 1, 1–9. *(Cited p. 235)*

[WVP$^+$94] L. Wehenkel, T. Van Cutsem, M. Pavella, Y. Jacquemart, B. Heilbronn, and P. Pruvot, *Machine learning, neural networks and statistical pattern recognition for voltage security: a comparative study*, Engineering Intelligent Systems for Electrical Engineering and Communications **2** (1994), no. 4, 233–245. *(Cited pp. 164, 218, 222)*

[WVRP86] L. Wehenkel, T. Van Cutsem, and M. Ribbens-Pavella, *Artificial intelligence applied to on-line transient stability assessment of electric power systems (short paper)*, Proc. of the 25th IEEE Conf. on Decision and Control (CDC), December 1986, pp. 649–650. *(Cited pp. 115, 129, 194)*

[WVRP87] L. Wehenkel, T. Van Cutsem, and M. Ribbens-Pavella, *Artificial intelligence applied to on-line transient stability assessment of electric power systems*, Proc. of the 10th IFAC World Congress, July 1987, pp. 308–313. *(Cited pp. 129, 194)*

[WVRP89a] L. Wehenkel, T. Van Cutsem, and M. Ribbens-Pavella, *An artificial intelligence framework for on-line transient stability assessment of power systems*, IEEE Trans. on Power Syst. **PWRS-4** (1989), 789–800. *(Cited pp. 3, 110, 115, 130)*

[WVRP89b] L. Wehenkel, Th. Van Cutsem, and M. Ribbens-Pavella, *Inductive inference applied to on-line transient stability assessment of electric power systems*, Automatica **25** (1989), no. 3, 445–451. *(Cited p. 194)*

[XRG+93] Y. Xue, P. Rousseaux, Z. Gao, L. Wehenkel, M. Pavella, R. Belhomme, E. Euxibie, and B. Heilbronn, *Dynamic extended equal area criterion - Part 1. Basic formulation*, Proc. of the Joint *IEEE-NTUA* International Power Conference APT, September 1993, pp. 889–895. *(Cited p. 200)*

[XWB+92] Y. Xue, L. Wehenkel, R. Belhomme, P. Rousseaux, M. Pavella, E. Euxibie, B. Heilbronn, and J.F. Lesigne, *Extended equal area criterion revisited*, IEEE Trans. on Power Syst. **PWRS-7** (1992), 1012–1022. *(Cited p. 200)*

[XZG+93] Y. Xue, Y. Zhang, Z. Gao, P. Rousseaux, L. Wehenkel, M. Pavella, M. Trotignon, A. Duchamp, and B. Heilbronn, *Dynamic extended equal area criterion - Part 2. Embedding fast valving and automatic voltage regulation*, Proc. of the Joint *IEEE-NTUA* International Power Conference APT, September 1993, pp. 896–900. *(Cited p. 200)*

[ZAD92] A. Zighed, J. P. Auray, and G. Duru, *Sipina. Méthode et logiciel*, Alexandre Lacassagne - Lyon, 1992. *(Cited p. 115)*

[Zur90] J. M. Zurada, *Introduction to artificial neural systems*, West Publishing, 1990. *(Cited p. 90)*

Index

Glossary

$3\phi SCs$: three-phase short-circuit, 203
CT : class probability tree, 103
DT : decision tree, 102
RT : regression tree, 102
H_m : maximal value of the entropy of a leaf, 135
KNN : K nearest neighbor method, 16
M : number of test states, 29
N : number of learning states, 28
P_e : error probability (and its test set estimate), 37
P_e^{Bayes} : error rate of the Bayes rule, 57
T : a tree, 101
α : risk of not detecting the statistical independence in stop-splitting, 109
m : number of classes, 27
n : number of attributes, 29
n : the number of candidate attributes, 26
$n_{i.}$: number of learning states of class i, 50
r : number of regression variables, 30
$\mathcal{N}_i$: an *interior* or test node, 101
$\mathcal{N}_t$: a *terminal* tree node, 101
$\mathcal{N}$: a tree node, 101
$\mathcal{N}_i$: set of all interior or test nodes, 101
$\mathcal{N}_t$: set of all *terminal* nodes of a tree, 101
$\mathcal{N}$: set of all nodes of a tree, 101
$\#\mathcal{N}$: total number of nodes of a tree, 119
LS : learning set, 28
PS : pruning set, 111
TS : test set, 29
U : universe of possible objects, 26

AE : absolute error, 139
AI : artificial intelligence, 99
AL : automatic learning, 1
ANN : artificial neural network, 11
AutoClass : clustering method, 69

BFGS : Broyden-Fletcher-Goldfarb-Shanno algorithm, 86

CCT : critical clearing time, 5
CPU : central processing unit, 10
CSVC : coordinated secondary voltage control, 208

DB : data base, 24
DM : data mining, 24
DSA : dynamic security assessment, 161
DT-ANN : Decision Tree - Artificial Neural Network, 150

EDF : Electricité de France, 95
ED : external disturbance, 176
EHV : extra-high voltage, 4
ERR : generalized error function, 77

FACTS : flexible alternating current transmission system, 168

GA : genetic algorithm, 155

HV : high voltage (93 and 63kV), 207

IBL : instance based learning, 100
ILP : inductive logic programming, 47

JAD : just after disturbance state, 94

KB : knowledge base, 162
KDD : knowledge discovery in data bases, 24

LPM : load power margin, 219
LS : learning set, 6
LTU : linear threshold unit, 72
LVQ : learning vector quantization, 97

MDL : mininum description length principle, 45
MH : modeling hypothesis, 177
MLP : multilayer perceptron, 13
ML : machine learning, 6, 99
MV : medium voltage (20kV or below), 207

NN : nearest neighbor method, 17

OLTC : on-load-tap-changers, 207
OMIB : one-machine-infinite-bus, 4
OP : operating point, 176
OS : operating state, 205

PR : pattern recognition, 2
PSA : probabilistic security assessment, 236
PSS : power system stabilizer, 249

RBF : radial basis functions, 89

SA : power systems security assessment, 2
SC : short-circuit, 203
SE : mean square error, 13
SMART : projection pursuit method, 57
SOM : self organizing feature map, 91
SVC : static VAr compensator, 168

TDIDT : top down induction of decision trees, 6
TS : test set, 6

WAMS : wide area monitoring system, 249